CHEM 107
Laboratory Manual

Experiments for
Essentials of General and Organic Chemistry

Tenth Edition

Renae R. Bullock (editor)
Liberty University

Experiments for Essentials of General and Organic Chemistry Laboratory Manual, Tenth Edition
CHEM 107
Copyright © 2018 by Liberty University

Scripture quotations marked (NKJV™) are taken from the New King James Version®. Copyright © 1982 by Thomas Nelson, Inc. Used by permission. All rights reserved.

Scripture quotations marked (ESV) are taken from The Holy Bible, English Standard Version® (ESV®), copyright © 2001 by Crossway, a publishing ministry of Good News Publishers. Used by permission. All rights reserved.

Scripture quotations marked (NASB®) are taken from the New American Standard Bible®, Copyright © 1960, 1962, 1963, 1968, 1971, 1972, 1973, 1975, 1977, 1995 by The Lockman Foundation. Used by permission. (www.Lockman.org)

Requests for permission to make copies of any part of the work should be mailed to:

Permissions Department
Academx Publishing Services, Inc.
P.O. Box 208
Sagamore Beach, MA 02562
http://www.academx.com

Printed in the United States of America

ISBN-13: 978-1-68284-420-5
ISBN-10: 1-68284-420-X

Contents

Acknowledgments

This work is dedicated to the glory and honor of my Lord and Savior, Jesus Christ, in whom "all things hold together" (Colossians 1:17 ESV).

I would like to thank and acknowledge my colleagues at Liberty University—Dr. Randy Davy, Dr. Mark Hemric, Dr. Nancy Richardson, Dr. Michael Korn, Dr. Alan Fulp, and Prof. Renae Bullock—for the contribution they made by sharing of their knowledge, for allowing me to edit some of the experiments they have written in the creation of this book, and for their valuable input and suggestions. Without your contribution this lab manual would not be the same, and your friendship means a lot to me personally!

A great thank you to my editor at Academx Publishing Services, Joann Manos, for her help in preparing this book. I also want to thank my past students in our General, Organic and Biochemistry course for pre-nursing majors, who have provided input and feedback for many of these experiments and encouraged me to put them in the form of a book—as well as my current and future CHEM 107 students, who will help correct any remaining mistakes. It is an honor and privilege to have an opportunity to teach you. This book is for you!

Next, I want to acknowledge Mr. Bill Childs and Mrs. Sue Sawyer, who were my mentors at Kellogg Community College and who still are my very good friends. Thank you both for your example in putting together your own lab manuals and for everything I have learned from you.

Finally, I owe a debt of gratitude to my father, Prof. Mark Goldin of Mendeleev University of Chemical Technology of Russia and N.V. Sklifosovsky Institute of Emergency Medicine in Moscow, Russia—for your inspiration to strive for excellence in all things; and an equally large debt of gratitude to my wonderful wife Lisa—for putting up with my long evenings working on this, for helping make this book as typo-free as it can be, and for all the support and encouragement you give me daily!

You shall love your neighbor as yourself.

– Matthew 22:39 (NKJV)

Safety Information

Safety is extremely important to working in the chemistry lab and is a part of every job you will do throughout your life. Legally and morally, individuals are obligated to take certain measures to protect themselves, others, property, and the environment. Teachers and employers are obligated to provide training about safety practices and laws and about how to be protected from hazards.

The federal government sets certain rules that apply to safety issues. Some are designed to protect workers (Title 29 of the Code of Federal Regulations (CFR), Section 1910 and specifically 1910.1450 for laboratories and 1910.1200 for hazard communication). The Occupational Safety and Health Administration (OSHA) enforces these rules. Other rules regulate waste disposal to protect the environment (Title 40 CFR) and are enforced by the Environmental Protection Agency (EPA). Some rules are spelled out in great detail, while others simply point to guidelines and have the expectation that "prudent practices" will be followed. Not following these prudent practices can result in severe penalties (like being charged in criminal court!), even when these practices are not fully written out in the regulations.

Since you are working in the laboratory this semester, you also must follow these safety rules! And this is not only to obey the government laws and rules... it's also to protect yourself and your classmates, in keeping with the teaching of Jesus in Matthew 22:39. However, we realize that it would be mighty hard for you to figure out all of these by yourself. Today's lab is designed to help!

You will receive some practical instruction in lab safety and general practices to be followed in all CHEM 107 laboratory experiments. These are based upon the written regulations and commonly accepted prudent practices. Following the rules will help keep you and your classmates safe! Also, avoiding accidents will save you time and help you learn more. We will keep you responsible for observing the rules that follow.

The rules given here are just a short summary of our "official" rules. You can access the more detailed safety documents on Blackboard, and we encourage you to read them after you have completed this lab activity. There is also a safety quiz that you must complete before the next experiment to be allowed to work in lab for the rest of the semester. This will count towards your lab grade as well!

RULES OF SAFE CONDUCT IN THE CHEMISTRY LABORATORY (abbreviated)

I. Protect yourself in the laboratory:

A. **Protect the eyes:** Wear approved eye protection AT ALL TIMES when glass, fire, or chemicals are present. If you wear eyeglasses, YOU MUST WEAR SIDE PROTECTION! Inform the instructor if you wear contacts. (While allowed, caution is advised if contacts are worn.)

B. **Dress appropriately**: Restrain long hair; avoid loose clothing and flammable materials. Wear shoes which cover the foot. Sandals are prohibited at all times.

C. **Keep floors, aisles, and benchtops clear** of any personal items such as backpacks, coats, etc. Keep chairs out of aisles as much as possible. This will prevent accidents such as trips, falls, or damage your belongings because of spills of chemicals or fires.

D. **Do not eat, drink, inhale fumes, or apply make-up**: Do not put anything into your mouth or onto your skin. This includes any food, drink, or medical inhalers, even on days when chemicals are not used. Wash your hands when you leave. Use proper ways to detect odors!

E. **Behave**: Avoid all kinds of practical jokes and horseplay in the lab. Never startle anyone. Never act silly or play in lab.

F. **Pay attention**: Read labels. Be aware of what is happening around you and of directions which have been given. Take notes of what is said in pre-lab lectures. Ask questions when you do not understand or when a label is unclear to you.

II. Know the location and operation of safety equipment:

A. **Fume hood**

B. **First aid kit**

C. **Eyewash**

D. **Safety shower**

E. **Fire extinguisher**

F. **Fire blanket**

G. **Exit doors** to the laboratory (some doors are locked with a "not an exit" sign)

H. **Emergency phone number** for LU police/emergency services: (434) 592-3911.

III. Know procedures for laboratory emergencies:

A. **Chemicals in the eyes**: flush eyes with water from the eyewash for 15 minutes.

B. **Corrosive chemicals on the skin**: flush with water in the sink. Use the safety shower for a large skin area when the sink is inadequate.

C. **Fire on a person**: help the person to **STOP, DROP, and ROLL!** Smother the fire any way possible, as fast as possible.

D. **Fire in a beaker**: cover with a watch glass to smother.

E. **Large fire**: Evacuate! Inform the instructor or other responsible person. If trained, use the fire extinguisher. Pull the pin and aim at the base of the fire. (This is very rare.)

F. **Minor injury**: inform instructor who will get you a bandage or whatever else you need.

G. **Serious injury**: inform the instructor, or, if instructor is absent, go to the office or look for another faculty member. If no one can be found immediately, call LUPD at (434) 592-3911.

IV. Know safe practices for equipment handling and chemical usage:

A. <u>**Spills:**</u> Clean up spills immediately, especially around or on balances and in common areas and whenever acids or bases are spilled. If a hazardous chemical is spilled, **do not attempt to clean up yourself!** Let the instructor know immediately, so the spill can be taken care of.

B. <u>**Glassware:**</u> Do not heat beakers to dryness—they break. Clean up broken glass immediately and place in box labeled "broken glass." Remember that *hot glass and cold glass look alike*!! Be careful what you touch.

C. <u>**Hand protection**</u>: Some chemicals are especially hazardous to your skin (such as acids, bases, and toxic substances). For these, the experiment procedure or the instructor may direct you to use gloves. Also, take care when working with hot objects—use a "hot mitt" or tongs to handle it instead of your hands.

D. <u>**Dispensing bottles:**</u>

    a. KEEP THEM WHERE YOU FIND THEM—other students need to use them too! If a chemical is kept under a fume hood, it may be unsafe to remove it from the hood.

    b. Use the scoops, spatulas, and droppers provided in the dispensing bottles; your own could contaminate the whole bottle.

    c. Do not put excess back into dispensing bottles *unless directed to do so by the instructor*. If possible, do not discard the excess, but give it to another student.

E. <u>**Chemical disposal**</u>: ALL USED CHEMICALS WILL BE COLLECTED BY THE INSTRUCTOR! STUDENTS **CANNOT** POUR ANY CHEMICALS DOWN THE DRAIN (except pure water)!

F. <u>**Questions**</u>: Ask questions when you are unsure of equipment operation or setup.

V. Be aware of hazards associated with materials you are using:

A **hazardous material** is anything that is flammable, corrosive, toxic, or highly reactive. However, just as healthcare workers take universal precautions to protect against pathogens even if a particular patient is not known to be a carrier of disease, we should treat all materials in science laboratories carefully, as if they were harmful – until we know that they are not. Such precautions include personal protection in the form of eye protection and gloves and care not to expose materials to flames.

Your experiments in this lab manual will provide all of the safety information you need to know for safely handling chemicals in lab each week. However, you can also find information about any chemical by consulting its **MSDS**, or **SDS**. "(M)SDS" stands for (Material) Safety Data Sheet. Anyone who works with chemicals must be trained on how to read these sheets and be provided access to them. Our laboratories have books of these sheets for each of the chemicals used in each lab. Many websites also have (M)SDS's. The fastest way to find a sheet is to go to Google and type the name of the material and "MSDS."

HOW TO READ SAFETY DATA SHEETS

Section 1, Identification: product name; manufacturer or distributor name, address, phone number; emergency phone number; recommended use; restrictions on use.

Section 2, Hazard(s) identification: all hazards regarding the chemical; required label elements.

Section 3, Composition/information on ingredients: chemical ingredients; trade secret claims.

Section 4, First-aid measures: important symptoms/effects, acute, delayed; required treatment.

Section 5, Fire-fighting measures: proper extinguishing techniques; chemical hazards from fire.

Section 6, Accidental release measures: emergency procedures; protective equipment; containment and cleanup.

Section 7, Handling and storage: precautions for safe handling and storage, incompatibilities.

Section 8, Exposure controls/personal protection: OSHA's exposure limits; exposure-reducing systems such as ventilation ("engineering controls"); personal protective equipment.

Section 9, Physical and chemical properties: description of the chemical's characteristics.

Section 10, Stability and reactivity: chemical stability and possibility of hazardous reactions.

Section 11, Toxicological information: includes routes of exposure; related symptoms, acute and chronic effects; numerical measures of toxicity.

Section 12, Ecological information

Section 13, Disposal considerations

Section 14, Transport information

Section 15, Regulatory information

Section 16, Other information

While all MSDS's will look slightly different, and older sheets may be organized differently, the information provided is essentially the same.

Background (for the Experimental Part)

Besides learning about safety rules, in today's experiment you will use some of the equipment found in your equipment drawer to begin learning proper ways to record measured numbers. Since chemistry is a **quantitative science,** chemists use measured numbers to numerically express such things as the quantity of a substance (such as mass or the number of particles), space taken up by it (volume), or the quantity of energy transferred to it or from it (such as heat).

Mass measurements are quite simple when a digital **balance** is used (see Figure 1). A balance is the chemists' preferred word for a scale used to weigh small quantities precisely For example, to determine the mass of the 10-mL graduated cylinder that should be found in your drawer, you would make sure that the balance is turned on and **zeroed** (that is, it shows a **reading** of zero on its screen, also called its **readout** when nothing is placed on the **pan** of the balance). You would then simply place the graduated cylinder onto the pan, wait until the reading on the readout stops changing, or nearly

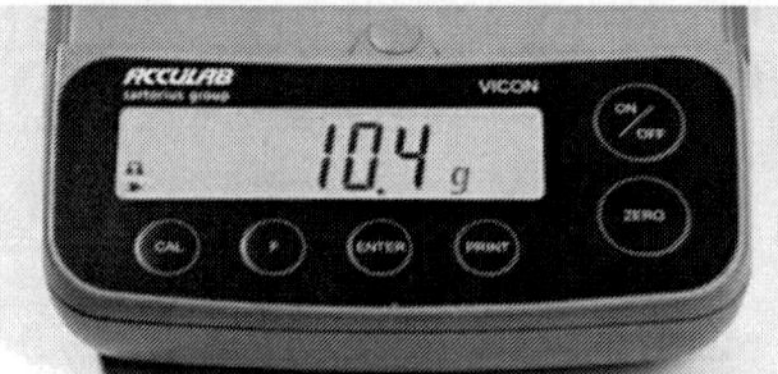

Figure 1. A digital readout

stops changing, and record the reading as your measured value of the cylinder's mass.

When an analog measuring tool, such as a ruler or a "dial" speedometer (with a needle) is used to record a measurement, it requires a little bit more thinking.

Figure 2. An analog measuring tool

The first step in recording a measurement on an analog measuring tool is to understand how to read the instrument. A metric ruler (shown in Figure 2) measures distance in centimeters (cm). There are ten graduations between each centimeter, meaning that the smallest graduation is 0.1 cm. 0.1 cm =1 millimeter (mm). So there are 10 mm in each cm. Manufacturers sometimes put "mm" on a metric ruler, even though the numbered graduations indicated centimeters.
A block of wood that is lined up exactly with the zero marking and extends to the halfway line between 4 cm and 5 cm,has a measurement of exactly 4.5 cm. This is the measurement because there are ten graduations between each centimeter.

However, if you look closely at Figure 2, you will see that the wooden block actually lines up somewhere between the markings for 4.5 cm and 4.6 cm. This creates a difficulty, since writing down that the length of the block is "between 4.5 and 4.6 cm" is an accurate observation, but an awkward and lengthy way of stating this fact.

Scientists have devised a recording convention for measured numbers. In fact, how a number is recorded (or specifically, the number of digits in a number) will be extremely important when the measured number is used in calculations; the rules that describe how measured numbers should be recorded and used in calculations are called the **rules of significant figures**.

To properly record the measurement according to the rules, start by recording all of the so-called **certain digits** (the ones that are indicated with the graduation marks). The last of these always is the same place as the smallest marking/graduation on the measuring tool. An additional digit is recorded by estimating where the end of the item falls between the two known graduation marks. This estimated digit is called an **uncertain digit**. In this case, the number is greater than 4.5 but definitely less than 4.6, so the certain digits would be **4.5** and the entire recorded number will be a **4.5__**, with the blank representing the uncertain digit to be estimated. Any number can be guessed for the uncertain digit, so one could record this measured number as **4.5<u>1</u>**, as **4.5<u>4</u>**, or as **4.5<u>8</u>** and not be wrong! Notice that the tenths place is also the smallest graduation marked on the ruler, so the hundredths place is the uncertain digit!

What is perhaps a bit harder to accept about the rules of significant figures is that, even though sometimes a measurement comes in "dead-on," like a dead-on 4.5 on the ruler in the first paragraph on this page, it still has to be recorded according to the same rules. In this case we know the digits **4.5** for sure – in other words, they are certain digits! However, we must still make an estimation on an uncertain digit, since the "5" is in the tenths place, and it actually corresponds to a graduation (marking) on the ruler. An excellent guess for a "dead-on" 4.5 would be recording it as **4.5<u>0</u>**, with the zero indicating that it is, indeed, dead on!

Graduated cylinders are used to measure the volume of liquids. The top part of the liquid in the graduated cylinder forms a U-shaped surface called a **meniscus**. *__To properly measure volume, record the reading at the lowest part of the meniscus while looking at it straight on, at eye level__* (see Figure 3). Otherwise, the reading will be too high or too low, depending on whether you are looking down or up at it.

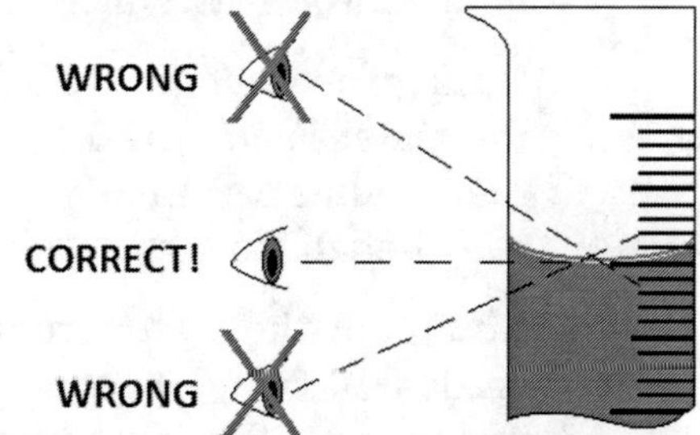

Figure 3. Proper volume measurement

Objectives

1. To learn safety procedures applicable to the chemistry lab

2. To learn/review names of commonly used lab items

3. To learn to record measured numbers with proper significant figures

4. To measure masses and volumes and record them using the rules of significant figures.

Procedure

Safety/Equipment

1. Answer the safety questions in Part I of your Report Sheet.

2. Identify equipment items shown on your Report Sheet (Part II) and write down their names and uses. *HINT:* Refer to the inside the front cover of this lab manual for help!

3. Sketch equipment items named in Part II on your Report Sheet. (Do your best, even if you "can't draw!") If in doubt, use the equipment chart on the front cover or ask a classmate or the instructor. All the equipment mentioned can be found in the drawers at your station or in numbered and labeled drawers around the lab room.

4. Familiarize yourself with the common chemistry glassware used in this lab and its location in the drawers of your station.

Measurements

For all measured numbers in this lab, the Data Sheet page will have three lines: one for the certain digits in that measurement, one for the estimated (guessed) uncertain digit, and one for the "finished product" – the measured number recorded according to rules of significant figures.

5. Measure the mass of the 10-mL graduated cylinder. Make sure that the cylinder has graduation marks to the 0.1-mL. If you aren't sure, check with the instructor. Record all of the digits shown on the digital display screen on the first line on number 9 on the Data Sheet page. On the remaining two lines, write which of the recorded digits are certain and which is uncertain.

6. Add tap water (from the faucet) to the graduated cylinder until it is about half-full. Place the 10-mL graduated cylinder now containing water back onto the balance. Record the combined mass of water and cylinder as indicated on the digital display screen of the balance, on the first line of number 10 on the Data Sheet page. On the remaining two lines of number 10, write which of the recorded digits are certain and which is uncertain.

7. Using proper technique as described in the Background section, note the volume reading of the water inside the 10-mL graduated cylinder, in milliliters. Record the measurement on the first line of number 11 on the Data Sheet. On the remaining two lines, write which of the recorded digits are certain and which is uncertain.

8. Now pour the water from your small 10-mL graduated cylinder into your large 100-mL graduated cylinder. Again, using proper technique, note the volume reading of the water inside the 100-mL graduated cylinder. Record the proper number of digits, as described in

the Background section, on the first line of number 12 on the Data Sheet page. On the remaining two lines, write which of the recorded digits are certain and which is uncertain.

9. Using the mathematical formula/equation given on the Data Sheet page, calculate the mass of water by subtracting properly recorded mass of the 10-mL graduated cylinder (Line 9, left-most column!) from the properly recorded mass of the 10-mL cylinder with water (Line 10, left-most column!). The answer should have the same number of digits after the decimal as the two numbers being subtracted… (More about this next week!)

Homework!

10. Remember to complete the safety quiz on Blackboard.

name   section   date

Report Sheet

Part I. Safety

Answer the following questions. Be sure to complete the room diagram, found on the bottom of this page, during lab.

1. When should a spill be cleaned up? Where does broken glass go?

2. Why should dispensing bottles be left where you find them and NOT taken to your own station?

3. List three types of information that you might find on an MSDS sheet besides the chemical name:

    (1)

    (2)

    (3)

4. Describe one thing not addressed in these questions, but included in the written information or in the lecture that you learned about safety that will be helpful to you.

5. Complete this diagram during the lab time in the lab room: Neatly indicate the location of safety equipment listed in the Safety Procedures, page 2, as well as any other items indicated by the lab instructor.

Front
of
Room   Back
of
Room

| name | section | date |

Part II. Laboratory Equipment

6. Write the names of the following laboratory equipment items. (*HINT:* Not all of them will be in your equipment drawer!)

______________   ______________   ______________   ______________

7. Sketch the following equipment items below. (Do your best even if you "can't draw!")

Evaporating dish	Erlenmeyer flask	Filtration flask	Graduated cylinder
Spatula	Test tube	Test tube tongs	Plastic pipette
Scoopula	Glass funnel	Buchner funnel	Glass rod and rubber policeman

Part III. Equipment Location and Familiarization

8. Each pair of students will be assigned a station. Each station has a set of labeled drawers and cabinets. The labels indicate the station numbers and the items that should be in each drawer and cabinet.

9. Below is a diagram of a station. Please label each drawer according to the numbers on the drawers at your station. Then on the diagram in the correct area, write the items that are found in each drawer based on the labels on the drawers.

10. If items are missing from the drawers, please obtain these items from a stock source.

name	section	date

Data Sheet

Mass and Volume Measurements

Always include the abbreviation for the scale that is used when recording a measurement. Example 57.65 mL.

The "measured number" in this column needs to be recorded according to the rules of significant figures! This means all the places indicated by the measurement graduation marks and one digit of estimation need to be recorded!

**Be sure to use a 10mL graduated cylinder that is further divided into 0.1 mL; not 0.2mL!*

	Measured Number*	Certain Digits	Uncertain Digit
EXAMPLE: Volume of rubbing alcohol, mL	_57.65 mL_	_57.6_	_5_
9. Mass of small 10mL graduated cylinder, g*	_______	_______	_______
10. Mass of small graduated cylinder with water, g	_______	_______	_______
Mass of water, g** (#9-#10)	_______		
11. Volume of water in small graduated cylinder, mL	_______	_______	_______
12. Volume of water in large graduated cylinder, mL	_______	_______	_______

***The mass of water on this line is calculated by subtracting two measured numbers. It needs to be recorded according to the rules of significant figures! In this case, it means that the result should have the same number of decimal places (digits after the decimal point) as the two measured masses, both of which ought to have the same number of decimal places.*

Mike Goldin (Liberty University)

Who has measured the waters in the hollow of His hand,
Measured heaven with a span
And calculated the dust of the earth in a measure?
Weighed the mountains in scales
And the hills in a balance?

– Isaiah 40:12 (NKJV)

Background

The FBI's Ten Most Wanted Fugitives list gives a physical description of each of the criminals who have the dubious "distinction" of making the roll. Among the characteristics listed for each one are: gender; date and place of birth; build, height, and weight; and hair and eye color. Some of these can change quite a bit, like weight or hair color. Others, like age and gender, are much harder to conceal. By listing all of these characteristics, the FBI is often able to apprehend these criminals because someone recognizes them despite their efforts to conceal their identities.

In the same way, we can recognize a substance based on its characteristics, or **properties**. A substance may have a unique color, odor, melting point, or boiling point. These properties do not depend on how much of the substance is there and are called **intensive properties**. Density also is an intensive property and can be used as an important clue of its identity. However, if you take a sample of your substance, it will also have properties that do depend on the quantity of the substance present, like the mass or volume of the particular sample you have. Just like our weight or hair style might change over time, mass and volume of different samples *of the same substance* will be different. Such properties are called **extensive properties**.

The **density** of a substance is the ratio of its mass per unit volume. It can be calculated by dividing the mass of the sample of your substance by its volume, as shown in Equation 1:

$$d = \frac{m}{V} \qquad\qquad [1]$$

where d is density, m is mass, and V is volume. While mass and volume do depend on the quantity of a substance (these are extensive properties), density as their ratio remains constant as long as the temperature doesn't change (and is therefore an intensive property). The units usually used for density are **grams per milliliter** (g/mL), or grams per cubic centimeter (g/cc or g/cm³), which is the same since 1 mL = 1 cm³ = 1 cc. Density changes with temperature because of the change in the sample volume, as materials usually expand when heated and contract when cooled. If the density of an object is less than the density of a liquid, it floats; if it's higher than that of the liquid, it sinks.

Density can be measured in several ways. Usually it is necessary to measure the mass and volume of the sample and then divide mass by volume. Mass measurement is typically pretty straightforward: just use the balance to weigh the sample or object and record it. However, to measure the mass of liquids, the liquid cannot be simply poured on top of the balance! Instead, first find the mass of the clean and dry container that will be used for the liquid sample and then weigh the container with the liquid. First weigh the container you're going to use for your liquid sample (which should be *clean and dry!*), then weigh the container with the liquid. Then subtract the mass of the empty container to get the mass of the liquid alone. This method is called **weighing by difference**.

As mentioned in Experiment 1, volume measurements can be a bit more complicated because of the U-shaped surface of the liquid; the **meniscus**. ***To properly measure volume, record the reading at the lowest part of the meniscus while looking at it straight on, at eye level*** (see Figure 1) to avoid a reading that is too high or too low.

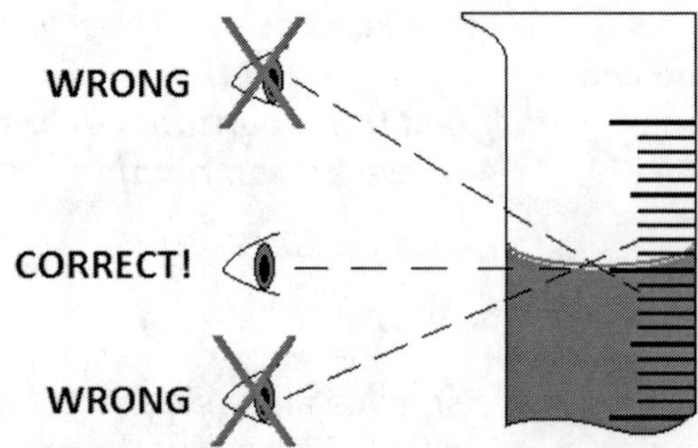

Figure 1. **Proper method for volume measurements**

If the object is a "regularly shaped" solid object, like a brick or a metal rod, you can measure its dimensions (length, width, and height; or radius and length), and then use a mathematical formula to calculate the volume. But there is no straightforward way to measure the size of "irregularly shaped" objects such as metal pellets or marbles! To measure "irregularly shaped" objects that don't dissolve or chemically react in water, a procedure called volume by displacement is used (see Figure 2). First, water is placed in a container and the volume recorded. Then the object is submerged in the water and the new water level is recorded. The difference between these two values is the volume of the object!

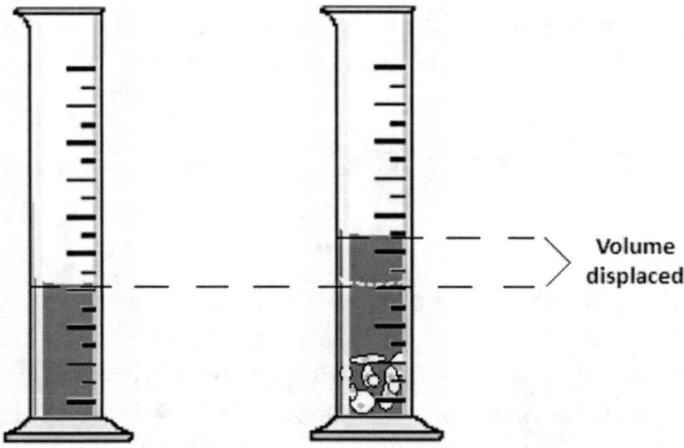

Figure 2. **Measurement of volume by displacement**

There is another way to measure density of liquids directly, and that is by using a **hydrometer** to determine the specific gravity of the liquid. **Specific gravity** is defined as the density of the substance divided by the density of water (Equation 2).

$$\text{Specific gravity} = \frac{d(\text{substance})}{d(\text{water})} \qquad [2]$$

Since the density of water is roughly 1.00 g/mL (as listed in Table 1), specific gravity is equal to the density of the substance **without units**, since g/mL gets cancelled when two densities are divided.

Significant Figures

> **IMPORTANT:** *You should be familiar with the rules of significant figures for recording measured numbers and performing calculations with measured numbers!*

To record mass, length, and volume measurements in this experiment:

- Record all known/certain digits from the ruler and graduated cylinder, then estimate the unknown digit. (*HINT:* You can use a "0" if your measurement is dead on a marking and a "5" if it's between markings.)

- For the digital readout on the balance, record **all** digits given. The last digit on the readout is the uncertain digit—the balance estimates this for you. *No need to estimate an additional digit!*

To calculate the volume of the regularly shaped metal block and cylinder, as well as the densities of each of the objects:

- For **addition/subtraction:** the result should have the same number of *decimal places* as the number with the fewest decimal places

- For **multiplication/division:** the result should have the same number of *significant figures* as the number with the fewest significant figures

- When both addition/subtraction *and* multiplication/division have to take place, follow the order of operations and figure out significant figures or decimal places at each step!

Densities and Identities of Substances

Since density is an intensive property, it can be used to help identify an unknown substance. However, knowing the density alone is not enough to be absolutely sure of what it is. It is necessary to know something else about it—another intensive property, such as its color, odor, melting or boiling point, or whether it's available in the lab.

In this experiment, you will identify several unknown solids and liquids. Assume that the substances listed in Table 1 below are available in lab and are your only choices for identification purposes. By checking your experimental value of density against the values in Table 1, you should be able to identify your unknown substances.

Table 1. Densities of some common substances

Solids	Density (g/mL)	Liquids	Density (g/mL)
Aluminum	2.70	Ethylene glycol	1.11
Copper	8.96	Gasoline	0.74
Cork	0.25	Glycerin	1.26
Glass	2.60	Hexane	0.66
Gold	19.3	Kerosene	0.82
Iron	7.87	Milk	~1.03
Lead	11.36	Olive oil	0.92
Magnesium	1.74	Rubbing alcohol	0.79
Tin	7.30	Sea water	1.03
Zinc	7.13	Water	1.00

Objectives

1. To measure masses and volumes and record them using the rules of significant figures.
2. To calculate the density of various substances based on mass and volume measurements.
3. To identify unknown substances based on their experimentally determined densities.

Materials and Equipment

You will need the following for your experiment:

1. Unknown metal block
2. Unknown metal cylinder
3. Unknown metal pellets or marbles
4. Unknown liquid
5. Metric ruler
6. 10-mL and 100-mL graduated cylinder
7. Top-loading electronic balance

Safety

- One of the unknown liquids is flammable and irritates skin and eyes on contact

Procedure

I. Density of a metal block

1. Obtain a solid block of an unknown metal. Using your metric ruler, measure the dimensions of the block (length, width, height) and record them with the correct number of significant figures.

2. Weigh the metal block. Record the mass with the correct number of significant figures.

3. Return the block to the instructor's desk, or its original location.

II. Density of a metal cylinder

4. Obtain a solid cylinder of an unknown metal. Record the letter listed on the container (Unknown A, B, or C).

5. Using your metric ruler, determine the dimensions of the cylinder (diameter, height) and record the values with the correct number of significant figures.

6. Weigh the cylinder. Record the mass with the correct number of significant figures.

7. Return the cylinder to the container at the instructor's desk, or its original location.

III. Density of a liquid

8. Weigh a clean, dry, empty 10-mL graduated cylinder. Use 10 mL cylinders that are further marked in 0.1 increments.

9. Measure about 7-8 mL of the unknown liquid into the graduated cylinder you weighed. Record the letter listed on the container (Liquid X or Y). Also record the *exact* volume with the correct number of significant figures.

10. Weigh the graduated cylinder with the unknown liquid and record its mass with the correct number of significant figures.

11. Discard about 2-3 mL of the unknown liquid. (***Use the correct discard collection container!***) Record the new volume *exactly* with the correct number of significant figures.

12. Now re-weigh the gradated cylinder. This is your second trial for the unknown liquid measurement!

13. Discard the rest of your liquid. (***Use the correct discard collection container!***)

IV. Density of pellets or marbles

14. Get some pellets of the unknown metal or several marbles:

    - If you will be using a 10-mL graduated cylinder for the volume measurement (below), you will want **about 6-8 metal pellets**.

    - If your pellets or marbles are too large to fit into a 10-mL graduated cylinder, use a 100-mL graduated cylinder. **In this case, use a much larger handful of pellets.**

    Record the number listed on the container (Unknown 1, 2, 3, or 4).

15. Weigh the pellets or marbles and record their mass with the correct number of significant figures.

16. Fill the 10-mL (or 100-mL) graduated cylinder approximately halfway with water. Record the volume of water in the cylinder with the correct number of significant figures.

17. Place the pellets or marbles into the graduated cylinder. Be sure all of the pieces are submerged. gently tap the sides of the cylinder with your fingers to ensure that no air bubbles are trapped in the metal. Record the new level of water in the cylinder with the correct number of significant figures.

18. Carefully catch the pellets or marbles by holding your hand over the mouth of the graduated cylinder as you drain the water into the sink. Dry them with a paper towel and lay them out to dry by the sink on a piece of paper towel.

Clean-Up

- Return all metal blocks and cylinders back to their original container.

- Lay out all the wet pellets or marbles on a paper towel to dry by either of the large sinks.

- Discard the unknown liquid (Liquid X or Y) into the correct designated collection container.

Calculations

*Your **experimental** density values may not match the values in Table 1 exactly. Use the closest value to your experimental density, as well as any additional information about the materials, to identify your unknown. Also, remember to use the rules of significant figures in all your calculations!*

I. Density of a metal block

1. Calculate the volume of the metal block using Equation 3:

$$V = w \times \ell \times h \qquad\qquad [3]$$

where *V* is volume (in cm³ = mL), *w* is width (cm), ℓ is length (cm), and *h* is height (cm).

2. Calculate the density of the metal block using Equation 1.

3. Using Table 1, determine the identity of the unknown metal.

II. Density of a metal cylinder

4. Copy the letter of the unknown metal cylinder from the Data sheet to the Calculations sheet.

5. Calculate the radius of the metal cylinder using Equation 4:

$$r = \frac{1}{2}\text{diameter} \qquad [4]$$

where r is the radius (cm) and d is the diameter (cm) of the cylinder. (*HINT*: Use the rule of significant figures for multiplication/division; remember that the "½" is exact!)

6. Calculate the volume of the metal cylinder using Equation 5:

$$V = \pi \times r^2 \times h \qquad [5]$$

where V is volume (in cm^3 = mL), π = 3.1415, r is radius (cm), and h is height (cm).

7. Calculate the density of the metal cylinder using Equation 1.

8. Using Table 1, determine the identity of the unknown metal.

III. Density of a liquid

9. Copy the letter of the unknown liquid (X or Y) from the Data sheet to the Calculations sheet.

10. Calculate the mass of the liquid by difference for Trial 1, as described in the Background section. Calculate the density of the liquid using Equation 1 for Trial 1.

11. Now calculate the mass of the liquid by difference for Trial 2, as described in the Background section. Calculate the density of the liquid for Trial 2.

12. Calculate the average density based on the densities of the two trials.

13. Using Table 1, determine the identity of the unknown liquid.

IV. Density of pellets or marbles

14. Copy the number of the unknown material from the Data sheet to the Calculations sheet.

15. Calculate the volume of the pellets or marbles by displacement, as described in the Background section.

16. Calculate the density of the metal pellets or marbles using Equation 1.

17. Using Table 1, determine the identity of the unknown material.

name   section   date

Pre-Laboratory Assignment

1. Define the following terms. You may use mathematical formulas instead of a definition or description.

    a. Density –

    b. Specific gravity –

    c. Weighing by difference –

    d. Measuring volume by displacement –

2. From the data in Table 1, which solid would float in water? *Explain.*

3. Using Table 1, determine the identity of an unknown liquid, if 25.00 mL of the liquid weigh 16.5 g. *Show your calculations. Use the correct rules for determining significant figures in calculations as shown in the Background section.*

4. A sample of metal pellets weighs 13.54 g. When this sample was placed in a 100-mL graduated cylinder, which contained 25.5 mL water, the new volume (of water and the pellets combined) was 27.0 mL. If the metal is one of the solids listed in Table 1, what is the identity of the metal? *Show your calculations and use the correct rules for determining significant figures in calculations as shown in the Background section. Explain the identification.*

____   __________   __________
name   section   date

Data Sheet

I. Density of a metal block:

      Dimensions: w, cm   ___________________________________

                          ℓ, cm   ___________________________________

                          h, cm   ___________________________________

      Mass, g   ___________________________________

II. Density of a metal cylinder: *Unknown ID* _______

      Dimensions: diam., cm   ___________________________________

                          h, cm   ___________________________________

      Mass, g   ___________________________________

III. Density of a liquid: *Unknown ID* _______

                                 Trial 1   *Trial 2*

      Mass of empty cylinder, g   _______________________   **(same as for Trial 1)**

      Volume, mL   _______________   _______________

      Mass of cylinder with liquid, g   _______________   _______________

IV. Density of pellets or marbles: *Unknown #* _______

      Mass of pellets/marbles, g   ___________________________________

      Volume of water, mL   ___________________________________

      Combined volume of water and pellets, mL   ___________________________________

name   section   date

Calculations

Show your work for each calculation! Use rules of significant figures; include units with each result.

I. Density of a metal block:

1. Calculate the volume of the metal block: $V =$ ___________________

    Show work:

2. Calculate the density of the metal block: $d =$ ___________________

    Show work:

3. What is the identity of the metal?   ___________________

II. Density of a metal cylinder: Unknown ID _______

4. Calculate the radius of the metal cylinder: $r =$ ___________________

    Show work:

5. Calculate the volume of the metal cylinder: $V =$ ___________________

    Show work:

6. Calculate the density of the metal cylinder: $d =$ ___________________

    Show work:

7. What is the identity of the metal?   ___________________

III. Density of a liquid: Unknown ID _______

| | *Trial 1* | *Trial 2* |

8. Calculate the mass of the liquid sample: $m =$ __________ $m =$ __________
    Show work:

9. Calculate the density of the liquid: $d =$ __________ $d =$ __________
    Show work:

10. Calculate the average density: $d_{avg} =$ ___________________
    Show work:

11. What is the identity of the liquid?   _______________________

IV. Density of pellets or marbles: Unknown # _______

12. Calculate the volume of the sample: $V =$ ___________________
    Show work:

13. Calculate the density of the material: $d =$ ___________________
    Show work:

14. What is the identity of the material?   _______________________

name   section   date

Post-Laboratory Assignment

1. In measurements with metal pellets (or marbles), if air bubbles were trapped among the pieces during volume measurement by displacement:

   a. Would the ***measured*** volume of the pellets be accurate, higher than the actual/true volume, or lower than the actual/true volume? *Explain.*

   b. Would the ***calculated*** density (based on measured mass and volume) be accurate, higher than actual/true density, or lower than actual/true density? *Explain.*

2. For the unknown liquid density measurement you performed:

   a. Were the densities from the two trials identical?

   b. Should they be identical? Why or why not? *Explain!* Hint: For this question, think in terms of intrinsic property of density. Refer back to the Background section if necessary.

3. Which graduated cylinder gives more **precise** measurements—the 100-mL one or the 10-mL one? *Explain.*

4. Are the ***measured*** mass, ***measured*** volume (by displacement) and ***calculated*** density of a wooden block accurate, higher, or lower than the actual/true value if...

    a.   the center is hollow, instead of solid, but the block is completely submerged (not floating)?

    Mass: _______________________________

    Volume: _______________________________

    Density: _______________________________

    HINT: In other words, how do the mass and volume of a hollow block compare to the mass and volume of a solid block? How will this difference in being hollow or solid affect the mass and then the calculated density?

    b.   the center is solid, but block is not totally underwater (is floating) when determining the volume by displacement?

    Mass: _______________________________

    Volume: _______________________________

    Density: _______________________________

    HINT: In other words, how does a solid block that isn't submerged compare in mass and volume to a solid block that is submerged? How does not being submerged affect the volume and then the calculated density?

5. You found a chest full of yellow coins in your grandma's attic. The coins have a mass of 2500 g and a volume of 221.4 mL. Are the coins gold? How do you know? Show your work and check that your results reflect proper significant figure rules.

EXPERIMENT 3: Calorimetry and Specific Heat

Mike Goldin (Liberty University)

To … the church in Laodicea write: … 'I know your works: you are neither <u>cold</u> nor <u>hot</u>.'

– Revelation 3:14-15 (ESV)

Background

Why does it take longer to boil a full kettle than just a cup of water? The answer is that it takes more energy to speed up the motion of water molecules when you have more of them in the full kettle. The type of energy associated with the motion of particles is called **heat**, or thermal energy. We observe heat when it *flows* (transfers) from one object to another. Heat always goes from an object at a high temperature to an object at a lower temperature (**from hot to cold**).

Temperature, therefore, is a *measure* of how much heat energy is contained in an object. It actually represents the average energy of all molecules in a sample. We usually measure it in degrees Fahrenheit (°F) or degrees Celsius (°C), although sometimes we must use the so-called "absolute" temperatures measured in kelvins (K)—in which zero is the lowest physically possible temperature. The first scientist who recognized the difference between heat and temperature was Joseph Black, a Scottish medical doctor; he is considered the "founding father" of **calorimetry**, the science of measuring heat. In fact, a type of calorimetry is still used in medicine today to measure how quickly heat is produced by cells in a culture or in a tissue specimen via metabolic processes.

Since heat flows from hot to cold, two objects at different temperatures that touch will reach the same temperature after enough time has passed. Once the two temperatures are equal, neither object is hot or cold relative to the other, so no more heat will flow; we then say that the objects are in **thermal equilibrium**. We can measure how much heat flowed from the hot to the cold object by using a device called a **calorimeter**. It is a container with insulated walls that prevents heat from coming into the calorimeter from the outside surroundings or going out into the sur-roundings. A perfect calorimeter would do this perfectly and would allow *no heat* to come in or leave through its walls. (Of course, no *real* calorimeter is "perfect," so some heat will flow between the calorimeter contents and the surroundings… but more about this later!)

Since we cannot *directly measure energy* (the heat that is transferred), we have to use a thermometer to measure the change in temperature of the calorimeter contents and then relate this change to the amount of heat that flows. A very simple calorimeter design, which we will use in this experiment, is shown in Figure 1.

In the example of boiling water in a kettle, the metal kettle will heat up very quickly, but it takes much longer for the water to heat up. Why do we observe this difference? The

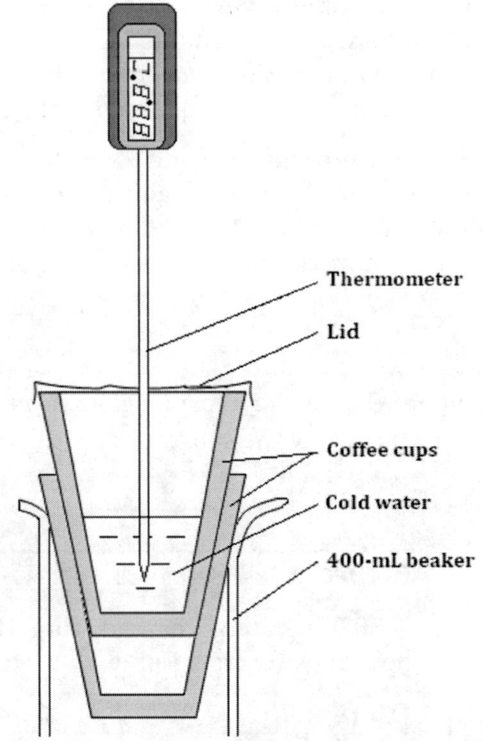

Figure 1. Coffee-cup calorimeter

answer to this is in the fact that *different substances differ in their ability to absorb heat*. Water can absorb a lot of heat with only a small rise in its temperature; but most metals only need a little bit of heat to greatly increase their temperature. We describe this ability of a substance to absorb heat as a *number* by measuring the quantity of heat (calories or joules) needed to raise the temperature of one gram of that substance by one degree Celsius. We call this property the **specific heat** of the substance. Specific heat is an intensive property and is measured in units of $\dfrac{J}{g\,°C}$ (joules per gram per degree Celsius) or $\dfrac{cal}{g\,°C}$ (calories per gram per degree Celsius).

Specific heats for some common substances are listed in Table 1. For example, the specific heat of water is 1.00 cal/g °C; this means that it would take 1 cal to raise the temperature of one gram of water by 1°C. In contrast, iron has a specific heat of 0.11 cal/g °C; it would take only 0.11 cal to raise the temperature of one gram of iron by 1°C.

<u>Table 1</u>. Specific heat values for some common substances

Substance	Specific heat (cal/g °C)	Substance	Specific heat (cal/g °C)
Lead	0.038	Glass	0.12
Tin	0.052	Table salt	0.21
Silver	0.056	Aluminum	0.22
Copper	0.092	Wood	0.42
Zinc	0.093	Ethyl alcohol	0.59
Iron	0.11	Water	1.00

We can use specific heats to glean different types of information. For example, among the *metals* listed in Table 1, we know that lead would be the quickest to rise in temperature if equal amounts of heat are transferred to equal masses of the metals; aluminum, on the other hand, is the slowest to heat up. Using the same reasoning, you could explain why a wooden building (such as a shed) doesn't get nearly as hot as a metal one would on a hot, sunny summer day.

In this experiment, you will learn how to obtain specific heat values from very simple measurements with a calorimeter. Hot metal will be added to water in the calorimeter, and its specific heat will be calculated based on the amount of heat transferred to the water. The heat required for a temperature change of a known amount (that is, mass) of a substance is given by the heat equation (Equation 1):

$$Q = m \times SH \times T \tag{1}$$

where Q is the heat lost or gained (in calories), m is the mass of the substance (in grams), SH is the specific heat of the substance (in cal/g °C) and T is the temperature change (in degrees Celsius):

$$T = T_{final} - T_{initial} \tag{2}$$

EXAMPLE 1

If 50.0 g water is heated from 10.0°C to 35.0°C, how much heat is needed? (Use the specific heat of water from Table 1.)

$Q = m \times SH \times T$

$$Q = 50.0\ \text{g} \times 1.00\ \frac{cal}{g\,°C} \times (35.0 - 10.0)\ °C = \boxed{1250\ \textbf{cal}}$$

So, 1250 calories of heat had to be transferred to (and absorbed by) by the water.

Energy can neither be created nor destroyed *by man* in any process but can be transferred from one object to another (the **Law of Conservation of Energy**). Therefore, the heat gained by water is exactly the amount of heat given up by the metal (as shown in Equation 3):

$$-Q_{metal} = Q_{water} \qquad\qquad [3]$$

The temperature of the water will rise as the temperature of the metal falls, until the two are equal (we would say that metal and water have reached **thermal equilibrium**). Since we know the specific heat of water (Table 1), the heat gained by water can be calculated, just as in Example 1.

If you use the specific heat of water (Table 1) and the measured mass and temperature change of water measured experimentally, you can calculate the heat gained by water, Q_{water}, and thus also the heat lost by the metal, Q_{metal}, from Equation 3. Then you can calculate the specific heat of the metal, SH_{metal}, by using the mass and temperature change of the metal measured in the experiment (as shown in Equation 4):

$$SH_{metal} = \frac{Q_{metal}}{m_{metal} \times \Delta T_{metal}} \qquad\qquad [4]$$

EXAMPLE 2

An unknown hot metal at 100.0°C with a mass of 50.03 g was placed into 40.11 g of water at a temperature of 21.5°C. A final temperature of 30.6°C was reached. The heat gained by the water is calculated by

$$Q_{water} = (40.11 \text{ g}) \times \left(\frac{1.00 \text{ cal}}{\text{g °C}}\right) \times (30.6 - 21.5)\,°C = 365 \text{ cal}$$

Heat lost by the metal is equal and opposite to heat gained by the water:

$$-Q_{metal} = Q_{water} = -365 \text{ cal}$$

The specific heat of the unknown metal is calculated:

$$SH_{metal} = \frac{-365 \text{ cal}}{(50.03 \text{ g})\times(30.6-100.0)°C} = 0.105 \text{ cal/g °C}$$

The specific heat of iron is 0.11 cal/g°C; thus, from the value of SH_{metal} determined experimentally, the unknown metal is iron.

As mentioned before, these calculations ignore the fact that the calorimeter itself absorbs part of the heat and then loses it to the surroundings—in other words, it's not a "perfect" calorimeter. To adjust for this, we could measure the **heat capacity** for the calorimeter—how many calories of heat are absorbed *by the calorimeter* each time the temperature goes up by 1°C, measured in $^{cal}/_{°C}$ (calories per degree Celsius). Measured heats would then be adjusted by this value. However, this is a fairly small error, and we will ignore it for purposes of this experiment. We will also have to use a special trick for measuring the final temperature of water, since it will start cooling down as soon as water reaches its maximum temperature because of the losses of heat through the walls of the calorimeter.

Objectives

1. To construct a simple calorimeter.

2. To measure the specific heat of a metal.

Materials and Equipment

You will need the following for your experiment:

1. Metal pellets
2. Styrofoam cups (2)
3. Plastic lids for the Styrofoam cups
4. Large test tube
5. Thermometers (2)
6. Graduated cylinder, 100-mL
7. Beakers: one small (50-150 mL), one 250-mL, one 400 mL

Safety Issues

1. Be careful as you work with hot objects! Do not touch hot glassware or hot plate surfaces!

2. Do not hand hot glassware to another person!

3. As always, wear your safety glasses, proper footwear, etc.

Procedure

1. Set up a **hot water bath** as demonstrated by the instructor. The hot water bath set up should include a stand and ring clamp around the beaker with hot water. This is to stabilize the beaker and prevent it from falling over and spilling boiling water. Fill a 600-mL beaker about halfway with "hot" water from the faucet. Set it on the hot plate and turn the "HEAT" knob to a setting of 300-350°C. Place an alcohol-based thermometer in the beaker to monitor the temperature of the hot water bath.

2. Place a dry, clean beaker on the balance and zero the balance. Weigh about 25-30 g of the metal pellets provided by your instructor in the beaker, and record the mass (with proper significant figures) on Line **2**. Pour the sample into a large, clean, dry test tube.

3. Clamp the test tube above the hot water bath and lower it in, so it is immersed in the hot water. The water in the hot water bath needs to remain above the level of the pellets in the test tube.

> *<u>IMPORTANT</u>: The bottom of the test tube should NOT touch the bottom of the beaker, and the water level in the beaker should always remain above the metal level in the test tube. **YOU SHOULD <u>NEVER</u> HAVE ANY WATER <u>INSIDE</u> THE TEST TUBE!!!***

4. While you are waiting for the water to begin boiling, construct a calorimeter, as shown in Figure 1. Place one dry Styrofoam cup into another and put the stacked cups into a 400-mL beaker (for stability).

5. In your 100-mL graduated cylinder, measure approximately 50 mL of cold water in the calorimeter cup. Since the density of water is nearly 1.00 g/mL over the temperature range for this experiment, recorded the actual number of milliliters in the graduated cylinder **with proper significant figures** as the mass, in grams, on Line **1** on the Data Sheet.

6. Cover the cup securely with a lid (or another cover, if provided by your instructor). The "coffee-cup" calorimeter is now ready for use. Use lab tape to cover any holes in the lid where heat could escape. Do not cover the hole where the thermometer will be inserted.

7. After the water *in the hot water bath* (in the beaker) starts gently boiling – reaches about 100°C and starts to bubble, start the timer and keep the metal in the hot water bath for 10 minutes. At the same time, begin to monitor and record the temperature of the cold water *in the calorimeter*: Stir the water inside the calorimeter (by swirling from time to time) and record temperature readings at 3-min intervals on the Data Sheet in the box under Item 4.

> **IMPORTANT: During this time, please read and be prepared for steps 8 and 9. These are very important steps and they happen quickly without much chance to think and read while they are occurring.**

8. After recording the water temperature reading in the calorimeter at 9 min, **the following steps must be done quickly and carefully:**

    a. record one more water/calorimeter temperature reading at 9.5 min (9 min 30 sec)

    b. at the same time, record the temperature reading on the thermometer in the hot water bath; record it on Line **3** (Data Sheet)

    c. remove the test tube from the boiling water

    d. dry the outside of the test tube with a paper towel to prevent dripping hot water in the calorimeter

    e. lift the lid with the thermometer off the calorimeter

    f. add the hot metal to the calorimeter as close to the 10-min mark on the timer as possible

    g. **immediately** reclose the calorimeter and begin to record temperature readings on the data sheet, starting with the line for 10.0 min (with an asterisk * beside it)

    Be careful no water is added to or lost from the calorimeter during the transfer.

9. Continue recording the calorimeter temperature on the Data Sheet, recording the temperature on the Data Sheet every 30 seconds for the next 4.5 minutes, until your timer reaches the 15-min mark. Stir the water before each time you record the temperature by gently swirling the calorimeter.

> **IMPORTANT: Do your best to avoid touching metal pieces inside the calorimeter with your thermometer because you need to be measuring the temperature of the *WATER*!**

Since you cannot see the metal pieces in the calorimeter, the best way to know whether the thermometer is touching the metal or the water is by looking at the direction of temperature change:

- If the temperature goes **UP FIRST, THEN DOWN**, it means that you are measuring the temperature of the water—so keep going!

- If the temperature is going **DOWN CONTINUALLY AND QUICKLY**, it means that you are measuring the temperature of the metal! To avoid the metal piece, move the thermometer around in the calorimeter and hold it a little higher, while still making sure it is submerged in the water. You may even need to momentarily lift the lid off and check to make sure the thermometer tip is in the water but not touching metal.

Clean-Up

- Pour out the water from the calorimeter inner cup, catching all the metal pieces into your hand.
- Lay out the metal pieces on a paper towel to dry by either of the large sinks.
- Use a piece of dry paper towel to dry the inside of the inner Styrofoam cup and lid.
- Return the Styrofoam cups, lids, and thermometers to the front desk.
- Use a piece of dry paper towel to dry any spilled water off your benchtop.
- Turn off and unplug your hot plate.

Calculations

1. Transfer the last temperature value of the cold water in the calorimeter before the hot metal pieces were added (found at the 9.5 min mark in the data table in Item 4 on your Data Sheet page) onto Line **5** of the Calculations sheet. This is the "initial" temperature of the water in the calorimeter.

2. Based on temperature data from the Data Sheet (the box under Item **4**), write down the "final" temperature of the water and the metal on Line **6** of the Calculations sheet.

> *__IMPORTANT__: The use of the terms "initial" and "final" here has spurred on many conversations and even arguments with students! This temperature is "initial" or "final" relative to the transfer of heat from the hot metal to the cold water.*
>
> *We only care about the conditions of the water in the calorimeter at two specific points:*
>
>   *1) when we began the transfer of heat from the hot metal to the cold water (that is, the moments immediately before the metal is first added)*
>
>   *2) when the transfer of heat from the metal to the water is over, and thermal equilibrium between the two is established.*
>
> *The water first <u>gains</u> heat when hot metal is added to it, and then it immediately starts <u>losing</u> heat because of the calorimeter's inability to totally prevent transfer of heat through the walls. So, we can tell approximately when the transfer of heat from the metal to the water ends – at the highest temperature of water during the time that its temperature of water inside the calorimeter was monitored and recorded.*
>
> *Therefore, the highest temperature value in the time/temperature table under Item 4 on your Data Sheet is considered to the "final" temperature of the water. And because of thermal equilibrium, this is also the "final" temperature of the metal. From your data, it is obvious that the temperature continues to go down after this for both water and metal – but it is <u>irrelevant</u> for our experiment and calculations!*

3. From the final temperature (Line **6** of the Calculations sheet) and initial water and metal temperatures (Lines **5** and **3** of the Data Sheet), calculate the temperature changes, ΔT, for the cold water and the metal (using Eq. 2). Write down your results on Line **7** and Line **8** of the Calculations sheet. *Be sure to show your work for all calculations!*

4. From the data on the mass (Line **1** of the Data Sheet) and temperature change (Line **7** of Calculations) of water, as well as the specific heat of water in Table 1, calculate the heat gained by water using Equation 1. Write it down on Line **9**. *Show your work!*

5. From heat gained by water, write down the heat lost by the metal (using Eq. 3) on Line **10**.

6. From the heat lost by the metal (Line **10**), the temperature change of the metal (Line **2**), and the data on the mass of the metal (Line **8**), calculate the specific heat of the metal using Equation 4. Write it down on Line **11**. *Show your work!*

7. Based on the specific heat of the metal (Line **11**), use Table 1 data to identify which of the metals from the table you were given. Write down your identification on Line **12**.

name	section	date

Pre-Laboratory Assignment

1. Define the following terms. You may use mathematical formulas instead of writing a definition or description.

    a. Calorimeter –

    b. Specific heat –

    c. Thermal equilibrium –

    d. Law of Conservation of Energy –

2. A student puts hot metal pellets into a container of cold water. Identify the direction of heat flow: from which object heat will flow and to which it will go.

3. You have a sample of iron and a sample of water, whose masses are equal. You increase the temperature of each sample by 10°C. Which material required more heat for the increase? Why? *HINT:* Refer to the concept of specific heat capacity in the Background section.

(continued on the back...)

4. Why should you not allow the large test tube with the metal pieces to touch the bottom of the hot water bath beaker? *Explain!*

5. Once the hot metal pellets are placed into the cold water inside the calorimeter, what can you do to make sure that the thermometer in the calorimeter is measuring the temperature of the water and not the metal pieces? (*HINT:* See the Procedure for details on this!) This is an important detail for correctly performing of the lab.

6. Which liquid would be a better coolant for a car engine, water or ethyl alcohol? Explain briefly. (*HINT:* A good coolant could absorb lots of heat and not go up very much in temperature. What relates ability to absorb heat with temperature? ... ☺)

name   section   date

Data Sheet

(1) Mass of the cold water, g (equal to its volume, mL)   ____________________

(2) Mass of metal, g   ____________________

(3) Temperature of hot metal (=temp. of boiling water), °C   ______ **100°C** ______

(4) Temperature of the water inside the calorimeter—use the spaces below:

Time (min)	Temperature (°C)	Time (min)	Temperature (°C)
0.0	____________	11.5	____________
3.0	____________	12.0	____________
6.0	____________	12.5	____________
9.0	____________	13.0	____________
* 9.5	____________	13.5	____________
** 10.0	____________	14.0	____________
10.5	____________	14.5	____________
11.0	____________	15.0	____________

*   This line is the last measurement of cold water *only* in the calorimeter, before metal is added

** Add the hot metal at this point; record the first combined temperature on this line

*** Highest temperature reached by calorimeter after addition of hot water or metal. Place 3 asterisks next to this value on the chart.

____   __________   __________
name   section   date

Calculations

Show your work for all calculations.

(5) Temperature of cold water and calorimeter, °C   __________________

(6) Equilibrium temperature of the water and metal***, °C   __________________

(7) *T* of the water, °C   __________________

   Show work:

(8) *T* of the metal, °C   __________________

   Show work:

(9) Heat gained by the water, cal   __________________

   Show work:

(10) Heat lost by the metal, cal   __________________

(11) Specific heat of the metal, cal/g °C   __________________

   Show work:

(12) Identity of the metal ____

*** The final temperature here is the highest temperature from data in the box under Item (4) on the previous page.

name   section   date

Post-Laboratory Assignment

1. How would the use of metal cups instead of styrofoam cups affect the heat measured—would Q_{water} increase, decrease, or stay the same? *Explain!*

2. A 30.0 g sample of water has heated from an initial temperature of 15.0°C to a final temperature of 50.5°C. How many calories has the water absorbed? *Show calculations and use correct significant figures.* *HINT:* First determine the ΔT.

3. A 50.0 g sample of metal beads was heated to 100.0°C. The metal was added to 25.0 g of water at a temperature of 21.0°C resulting in a temperature rise for the water to 40.0°C. *Show all work and use correct significant figures.*

   a. What was the temperature of the metal at equilibrium? *HINT:* Refer to the Procedures and Data Sheet for information on equilibrium temperature.

   b. What was the temperature change of the metal?

   c. What is the specific heat of the metal?

Mike Goldin (Liberty University)

He is before all things, and in Him all things hold together.
– Colossians 1:17 (NASB)

Background

The number of electrons in an atom is equal to the number of protons (called the **atomic number** of the element), since atoms are electrically neutral; this way, the number of + charges is equal to the number of – charges. Electrons in the atom can only have certain energy levels. This has to do with how the nucleus interacts with the electrons, and an exact explanation involves pretty complicated math! We call these energy levels electron **shells**. The higher the energy of a shell, the farther the electrons are from the nucleus (Figure 1).

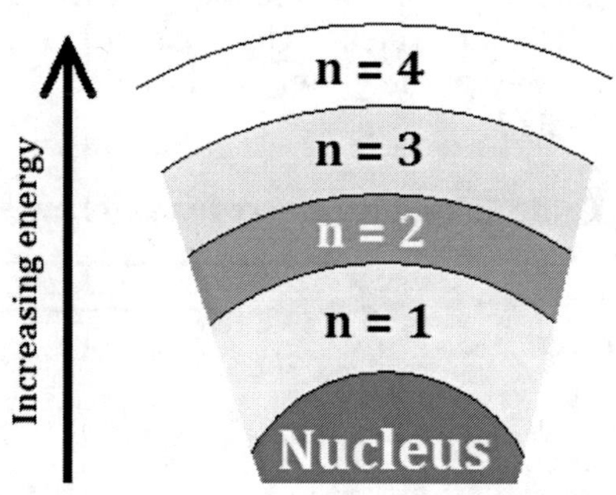

Figure 1. Electron energy levels (shells)

There is a limit to how many electrons can fit in a particular shell. If we use the letter n for the number of the shell (with $n = 1$ being the lowest-energy shell), the number of electrons that can fit in the shell can be calculated according to Equation 1:

$$\textbf{Maximum \# electrons} \text{ (in shell } n) = \mathbf{2n^2} \qquad [1]$$

Within the shells, electrons are located in regions of space called **orbitals**. The most important types of orbitals are the sphere-shaped s orbitals and "dumbbell"-shaped p orbitals, seen in Figure 2 below. There are also d orbitals and f orbitals that will show up in higher shells; d orbitals look a bit more complicated (only one of them is shown in Figure 2), and f orbitals (not shown) look even stranger!

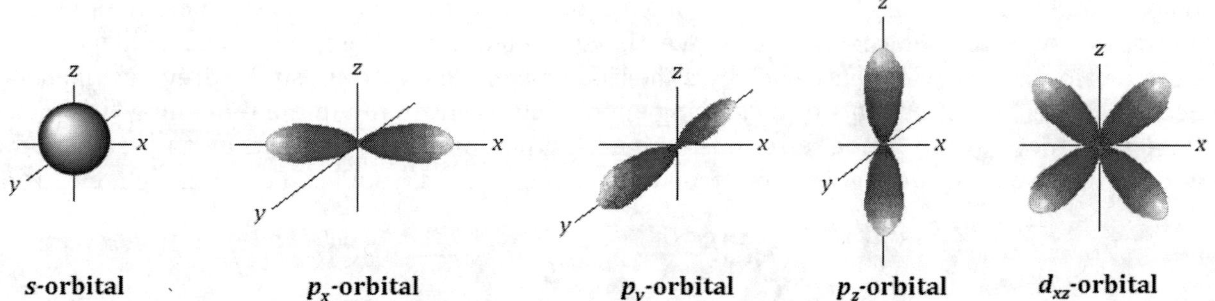

Figure 2. s, p, and d orbitals

There is one s orbital and three p orbitals in every shell (except the first one, which only has one s orbital, since it can only hold up to 2 electrons). d orbitals start in shell $n = 3$, and f orbitals, in $n = 4$.

Valence electrons

The most important electrons of an atom are the outermost electrons in the highest energy level (shell), since they are the ones that interact with other atoms. The outermost shell is called the

valence shell of the atom, and the orbitals and electrons in it are called **valence orbitals** and **valence electrons**. All properties of an element are determined by its number of valence electrons.

Because of some differences in energy between orbitals in higher shells, <u>**valence electrons are ONLY the electrons in *s* and *p* orbitals.**</u> Since there are one *s* and three *p* orbitals in every shell except the first one, there can be **up to eight valence electrons** in the four valence orbitals of any atom. We can represent the valence electrons as **dots** drawn around the element symbols on *four sides*, representing the four valence orbitals. This is called the **Lewis symbol** of the element. The number of valence electrons in the atoms of *representative* (A-group) elements is equal to the number of the group in which they are found in the Periodic Table. Some examples are shown in Table 1:

Table 1. **Valence electrons and Lewis symbols of some elements**

Element	Group	# valence e⁻	Lewis symbol
Sodium	IA	1	Na •
Chlorine	VIIA	7	$:\ddot{Cl}•$
Neon	VIIIA	8	$:\ddot{Ne}:$

The lowest possible energy state for an atom is to have a filled valence shell (eight electrons)—like neon (Table 1). Atoms will lose, gain, or share electrons to get eight of them in their valence shell; this observation is called the **octet rule** (from the Greek "oct," for "eight"). There are a few exceptions to this rule: hydrogen is the most important one—it only has room for two electrons in its valence shell and is said to follow a **duet rule**. There are other exceptions where atoms have more than eight valence electrons, but we will not concern ourselves with these in our workshop.

Ionic Bonds

When atoms gain or lose electrons, they form **ions**, called **cations** if they are positive (have lost electrons) or **anions** if they are negative (have gained extra electrons). Ions will have eight electrons in their "new" valence shell by either losing enough electrons, so the previous filled shell becomes their outermost one; or by gaining enough electrons to fill up the remaining "vacancies." Obviously, they have to "trade" electrons with each other, one atom giving its electrons to another. For example, sodium could give an electron to chlorine, so both would end up with eight electrons:

$$Na• \quad :\ddot{Cl}: \longrightarrow Na^+ [:\ddot{Cl}:]^- \longrightarrow NaCl$$

As a result, you get two ions, Na⁺ (sodium ion) and Cl⁻ (chloride ion), which become attracted to each other. The attraction force between oppositely charged ions that brings them together is called an **ionic bond.** Note that the Lewis symbols for the ions have their charges on them. You often see [brackets] around the Lewis symbols for ions. Na⁺ does have eight valence electrons, even though they are not shown; this is to show that Na• has *lost* electrons. <u>**Ionic bonds generally form between a metal and nonmetal.**</u>

Covalent Bonds

When two atoms share electrons to "gain" additional electrons and achieve octet, the shared pair(s) of electrons are called a **covalent bond. <u>Covalent bonds generally form between two nonmetal atoms.</u>** For example, two chlorine atoms could each share an electron:

$$:\ddot{\text{Cl}}\cdot \quad :\ddot{\text{Cl}}: \longrightarrow :\ddot{\text{Cl}}:\ddot{\text{Cl}}: \ (or\ :\ddot{\text{Cl}}-\ddot{\text{Cl}}:) \longrightarrow Cl_2$$

For oxygen, which needs to gain two electrons to get an octet, two pairs of electrons would be shared:

$$:\dot{\ddot{\text{O}}}\cdot \quad \cdot\dot{\ddot{\text{O}}}: \longrightarrow :\text{O}::\text{O}: \ (or\ :\text{O}=\text{O}:) \longrightarrow O_2$$

This type of covalent bond, where two pairs of electrons are shared, is called a **double bond,** and the covalent bond between chlorines above—a **single bond.** You can also have **triple bonds,** where three pairs of electrons are shared. (You will see one of these in this workshop.)

The types of pictures we have drawn for Cl_2 and O_2, are called **Lewis structures.** They show the bonding patterns in a molecule, as well as the electrons involved in them. Bonds (also called **bond pairs** of electrons) are usually shown with lines (—), and pairs of dots (:) represent unshared pairs of electrons, called **lone pairs.** Some of the covalent bonding patterns are very easy to figure out; others are not as obvious. Instead of trying to "piece together" Lewis structures of molecules, it is easier to follow a set of rules that achieve the correct structure through a different approach— by taking all of the valence electrons off all atoms and then redistributing them in such a way that every atom gets eight valence electrons.

Rules for Drawing Lewis Structures

1. **Add up the total # valence e– in molecule.**
   This is your "electron bank." Keep track as you make "withdrawals" and never "overdraw" the bank!

2. **Put "neediest" atom at center and connect all outer atoms to it with single bonds.**
   The atom that *needs to gain the greatest amount of electrons* always goes in the center. All other atoms connect to it with bonds. Subtract the electrons used up from the "bank."

3. **Put electrons in pairs on the outside atoms until each one has an octet.**
   Place pairs of dots on the outside atoms, until each one ends up with eight. The bonding electron pairs will count as part of the total electrons for each of the outer atoms. Subtract the number of electrons used up from the "bank."

4. **Put electrons in pairs on the central atom until it has an octet.**
   Place any remaining pairs of dots on the central atom until it ends up with eight. The bonding electron pairs will count as part of the total electrons for each of the outer atoms. Subtract the number of electrons used up from the "bank." *Sometimes there won't be any electrons left at this point.*

5. **Check whether all atoms now have an octet! If not, make multiple bonds** between the outside atoms and the central atom by *taking away lone pairs of electrons* from outside atoms and adding them as bond pairs for double, triple, etc., bonds. *NOTE: Sometimes more than one way of doing this is possible, ending up with different but equally correct Lewis structures.*

Shapes of Molecules

Bonds and lone pairs are negatively charged *electron pairs* that repel each other (because like charges repel). Therefore, the shape of any molecule with three or more atoms will be such that the bonds and lone pairs **on the central atom** are as far away from all the other bonds/lone pairs as possible. This explanation is called **valence shell electron pair repulsion** (or **VSEPR**). To figure out the shape of a molecule, you will use its Lewis structure to find the number of bonds and lone pairs around the central atom. (*Obviously, if only two atoms are there, there is no central atom, the only bond has nothing to repel from, and the only possible shape is linear!*)

Table 2 below shows the shapes of molecules with two, three, or four bonds and/or lone pairs on the central atom. Please note that the "drop"-shaped areas with a pair of dots in the drawings of the molecule shapes represent lone pairs; since they do not connect to an atom, they are "invisible" and not included in the name of the shape (like "trigonal pyramidal" or "bent"). The 2-dimensional representations in the column on the far right are how chemists *typically* draw molecule shapes. The wedge represents a bond coming out of the page, and the dashed line, going into the page.

Table 2. Molecule shapes according to VSEPR

e- Areas Around Central Atom	Bonds	Lone Pairs	Shape Name	3-D Picture of the Shape	2-D Representation
2	2	0	**Linear**		B—A—B
3	3	0	**Trigonal planar**	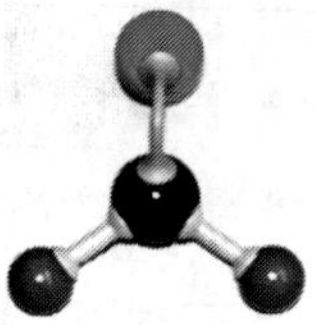	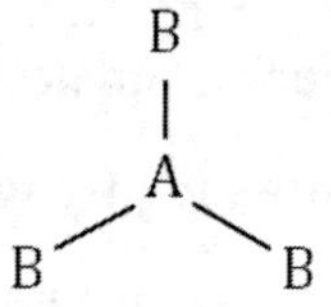
3	2	1	**Bent**	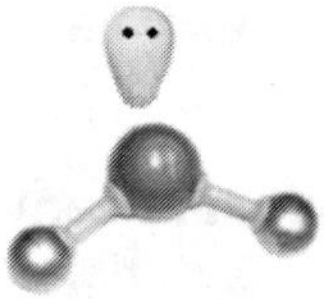	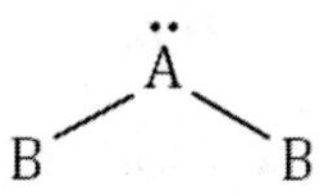
4	4	0	**Tetrahedral**	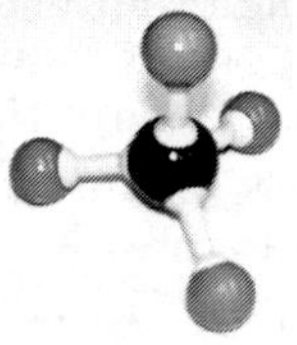	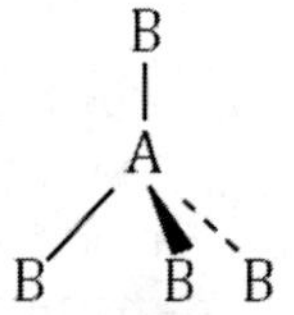
4	3	1	**Trigonal pyramidal**	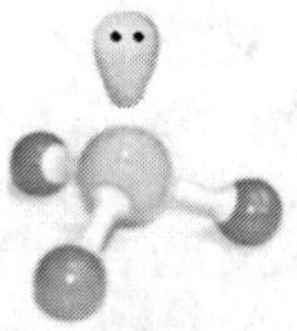	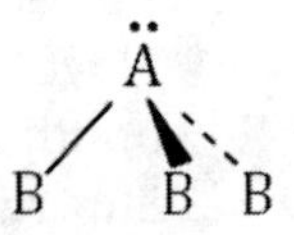
4	2	2	**Bent**	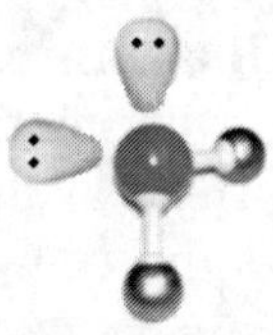	

It is possible to construct models of molecules that would look like the pictures in Table 2. We will use molecular model kits during this workshop to help you visualize the molecules.

Objectives

1. Predict whether two atoms will form an ionic bond or a covalent bond.

2. Develop skills in, and conceptual understanding of, Lewis symbols and structures:
   - Write Lewis symbols of elements and ions;
   - Write chemical formulas of ionic compounds;
   - Write Lewis structures of covalent compounds;
   - Predict whether two atoms will form a covalent or ionic compound.

3. Predict empirical formulas of ionic compounds and covalent compounds.

Materials and Equipment

You will need the following for your experiment:

1. A molecular model set

Safety

Even though this is a "paper experiment," chemicals may still be present on surfaces and inside the room. You must still:

- Wear proper clothing and shoes to protect yourself from the usual lab hazards.

- Refrain from eating or drinking anything in lab, as well as applying make-up.

- ***However, safety glasses or goggles will not be needed for this lab session!***

Procedure

Write Lewis symbols and Lewis structures of compounds shown on your Worksheet. For some of them, you will also predict their shape and put together molecular models.

Clean-Up

- Break up any molecular models you put together during the workshop.

- Put the "atoms" (balls) and "bonds" (sticks) in appropriate compartments in the box and close the lid.

- Brush any eraser shavings or paper shreds off the benchtops.

name　　　　　　　　　　　　　　　　　　　　section　　　　　date

Pre-Laboratory Assignment

1. Define the following terms, as they relate to this workshop:

    a.　　Valence electrons –

    b.　　Octet rule –

    c.　　Ionic bond –

    d.　　Covalent bond –

    e.　　Lone pair –

    f.　　Double bond –

2. Give the element symbol, group number, valence electrons, and draw Lewis symbols for the atoms (***not*** ions!) below:

Element	Element symbol	Group	# valence e^-	Lewis symbol of the atom
Potassium	**K**	**IA**	**1**	**K ·**
Carbon				
Phosphorus				
Bromine				
Argon				
Magnesium				
Aluminum				

3. For the elements below, give the number of valence electrons, how many electrons will need to be lost or gained by each to gain octet, whether an anion or cation is formed, and the Lewis symbol and charge of the ion (***not*** atom!).

Element	# valence e-	Lose/Gain e-?	Anion/Cation?	Lewis symbol
Sodium	1	lose 1	cation	Na^+
Oxygen	6	gain 2	anion	$[:\ddot{O}:]^{2-}$
Chlorine				
Aluminum				
Calcium				
Nitrogen				

4. Based on your answers to Question 2 above, give the chemical formulas of the ionic compounds formed by:

   a. Sodium and nitrogen

   b. Aluminum and oxygen

   c. Calcium and chlorine

5. Determine whether these substances have a covalent or ionic bond. (Write "**C**" for covalent or "**I**" for ionic on the blank line next to each.)

                                        Covalent/Ionic?

   a. Potassium chloride, KCl   ____________

   b. Hydrogen chloride, HCl   ____________

   c. Chlorine, Cl_2 ____________

   d. Sodium sulfide, Na_2S ____________

name				section	date

Worksheet

#	Formula	Total Valence e−	Lewis Structure	On **central atom**: Bonds	On **central atom**: Lone pairs	Shape (name and 2D drawing)
1.	I_2					
2.	N_2					
3.	H_2O					
4.	NH_3					
5.	CO_2					
6.	HCN					
7.	$CHCl_3$					
8.	OH^-	**8**				
9.	CN^-					
10.	CO_3^{2-}					
11.	NH_4^+	**8**				

name   section   date

Post-Laboratory Assignment

For each of the molecules on the Worksheet, determine their polarity and type of attractive forces between them. (Read sections 3.2-3.3 in Chapter 3 in the textbook to learn about bond and molecule polarity, as well as attractive forces; pp.90-103).

#	Formula	ΔEN	Bond is polar?	Shape (from Worksheet)	Molecule (or ion) is polar?	Attractive forces
1.	I_2					
2.	N_2					
3.	H_2O	1.4	YES	Bent	YES	Hydrogen bonds
4.	NH_3					
5.	CO_2					
6.	HCN	C-H / C-N	C-H / C-N			
7.	$CHCl_3$	C-H / C-Cl	C-H / C-Cl			

EXPERIMENT 5: **Chemical Reactions and Equations**

Mike Goldin (Liberty University)

... [T]here shall be equal amounts of each.
– Exodus 30:34c (NKJV)

Background

In our alcohol distillation experiment (Exp. 9), we will be using a difference in the physical properties of ethanol and water to separate the two. The actual separation is accomplished by **physical changes**, such as evaporation and condensation. These generally do not involve a change in the chemical bonds that hold together the individual ethanol or water molecules, and the identity of the substances did not change throughout the experiment—that is, we recovered the ethanol that was dissolved in mouthwash instead of making some other substance(s) out of it.

On the other hand, a change that involves breaking existing chemical bonds and making new ones in a different pattern is called a **chemical change**, or a **chemical reaction**. For example, burning natural gas (in a stove burner or a Bunsen burner) is a chemical reaction. Two substances, natural gas (methane, CH_4) and oxygen gas (O_2), **react**, and two new substances, carbon dioxide (CO_2) and water (H_2O), are produced as a result. We can describe this process by writing a **chemical equation** for this reaction in the following way (Eq. 1):

$$CH_4 + O_2 \rightarrow CO_2 + H_2O \qquad [1]$$

Notice that this equation resembles a mathematical equation: you have addition signs on the left and right side, and an arrow $\rightarrow$ is roughly equivalent to an equal sign =. The left-to-right direction of the arrow indicates that the chemical formulas on the left-hand side are the starting substances, called **reactants**, and the ones to the right of the arrow are the substances produced in the chemical change, called **products**. The bonds between carbon and the four hydrogens have been broken, as was the bond between the two oxygens in O_2, and new bonds are formed between carbon and oxygen in CO_2 and hydrogen and oxygen in H_2O.

There is one more aspect of chemical equations, however: to truly be "equations," their left and right side have to be "equal" in some fashion, just like in algebraic equations! What must be considered is the Law of Conservation of Mass, which states that matter cannot be created or destroyed (by man). For chemistry, this means that however many atoms of each element are there in the reactants must still be there in the products.

Let us look back at Equation 1: If we were to simply count the number of atoms of each element (C, H, and O) on the left side and the right side, we would quickly discover that, while both sides have one atom of C, there are four H and two O on the left and two H and three O on the right side. In other words, it *appears* from Equation 1 that two hydrogens were "destroyed" and one oxygen was "created" in that reaction. Of course, this cannot be! Therefore, we have to adjust a couple of things in our equation: if we take two molecules of O_2 and two molecules of H_2O instead of one, we would get this (Eq. 2):

$$CH_4 + 2\,O_2 \rightarrow CO_2 + 2\,H_2O \qquad [2]$$

What this means is that each molecule of CH_4 has to react with two molecules of O_2 to produce one molecule of CO_2 and two molecules of H_2O. It is not hard to see that now both the left and right side of the equation now have one C, four O, and four H atoms. While we will not discuss at this point the exact technique for determining how to make these adjustments, the numbers of molecules in

Equation 2 are called **coefficients**, and the adjusted equation is called a **balanced equation**. The original equation (Eq. 1) would be called an **unbalanced equation** to emphasize that it does not have all of the information to completely describe the chemical reaction. While we will not discuss the technique for balancing equations here in detail, it is not an overly difficult task!

In this experiment, we will take a closer look at the different patterns that most chemical reactions fall into, observe different types of reactions, balance chemical equations, and complete some chemical reactions if the reactants are known.

Types of Reactions

There are different ways we can split up all chemical reactions into different categories (types or classes). One of the most commonly used descriptions of reactions has to do with the pattern of what happens to the reactants. There are four main types of reactions according to this system. The first one is **combination** reactions (a.k.a. **synthesis**), where two or more simple reactants combine, resulting in one product. The overall pattern for combination reactions is shown in Equation 3 below, and Equations 4 and 5 are specific examples—the formation of ammonia (NH_3) from nitrogen (N_2) and hydrogen (H_2) gases, and the formation of carbonic acid (H_2CO_3) from carbon dioxide (CO_2) and water (H_2O):

$$A + B \rightarrow AB \tag{3}$$

$$3\ H_2(g) + N_2(g) \rightarrow 2\ NH_3(g) \tag{4}$$

$$CO_2(g) + H_2O(\ell) \rightarrow H_2CO_3(aq) \tag{5}$$

Notice that Eq. 4 has two *elements* as reactants, and Eq. 5 has two *compounds* as reactants—but both have a more complex, larger substance as products! By the way, notice also that Eq. 4-5 contain not only the chemical formulas and coefficients of the reactants and products, but also their physical states in parentheses. The accepted abbreviations for physical states are as follows:

- (g) = gaseous state
- (ℓ) = liquid state [notice that the "ℓ" is a cursive one instead of a printed "l"]
- (s) = solid state
- (aq) = aqueous [meaning the substance is dissolved in water and is a solute]

We must usually indicate the physical state of each substance in the chemical equation, since it is very important information for someone else on how the reaction was done.

Going back to the types of reactions, the second type is **decomposition** reactions. If you look at Equation 6, showing the general pattern of decomposition reactions, you will see that it is exactly opposite of the combination reactions: we begin with a more complex reactant, which breaks down (or **decomposes**) into simpler substances. The overall pattern for decomposition is shown in Equation 6 below, and Equations 7-8 are specific examples—the decomposition of water (H_2O) into hydrogen (H_2) and oxygen (O_2) gases under the influence of electrical current, and the decomposition of calcium carbonate ($CaCO_3$) into carbon dioxide (CO_2) and calcium oxide (CaO):

$$AB \rightarrow A + B \tag{6}$$

$$2\ H_2O(\ell) \xrightarrow{\textit{electricity}} 2\ H_2(g) + O_2(g) \tag{7}$$

$$CaCO_3(s) \xrightarrow{\Delta} CaO(s) + CO_2(g) \tag{8}$$

Notice that in Eq. 7 and 8, there are things written above the equation arrow—the word "electricity" in Eq. 7 and a Greek delta "Δ" in Eq. 8, which represents that heating is required for the reaction to happen. When words or symbols appear above (or, sometimes, below) the reaction arrow, it indicates the *conditions* necessary for the reaction to take place as written. You may see a specific temperature (for example, "500°C") required for the reaction, or a chemical formula (for example, "H_2O") indicating that the presence of another substance is required. This information is also crucial, so someone else can repeat our experiment.

The third type of reaction is called **single displacement** (a.k.a. **single replacement** or **substitution**) and involves the replacement of an element in a compound with another element, as shown in Equation 9. An example of this type of reaction is given in Equation 10, where aluminum, present as a *free element* (meaning in its elemental form, not bonded to another element in a compound), displaces hydrogen from nitric acid (HNO_3), which then becomes a free element (H_2) as a product of the reaction, along with aluminum nitrate ($Al(NO_3)_3$) as the other product:

$$A + BC \rightarrow B + AC \qquad [9]$$

$$2\,Al(s) + 6\,HNO_3(aq) \rightarrow 2\,Al(NO_3)_3(aq) + 3\,H_2(g) \qquad [10]$$

Finally, the fourth type of reaction is **double displacement** (a.k.a. **double replacement** or **metathesis**), which involves swapping of ions between two ionic compounds dissolved in water. Equation 11 shows the general pattern of this type of reaction, and Equation 12, an example of it, where lead(II) nitrate ($Pb(NO_3)_2$) reacts with potassium chloride (KCl) and the cations "switch partners," producing lead(II) chloride ($PbCl_2$) and potassium nitrate (KNO_3):

$$AB + CD \rightarrow AD + CB \qquad [11]$$

$$Pb(NO_3)_2(aq) + 2\,KCl(aq) \rightarrow PbCl_2(s) + 2\,KNO_3(aq) \qquad [12]$$

How do we know whether a chemical reaction will actually happen? Most of the time, you have to "just know" that it happens—in other words, from other chemists' experience of trying and succeeding at carrying out a particular reaction. There are certainly patterns we could discuss and rules we could derive, although we will not concern ourselves with this. One type of reactions can be easily explained—the double displacement reactions; however, we have to delve deeper into how it occurs first. To do this, we must discuss the behavior of ionic compounds in water.

Aqueous Solutions

In general, water is a nearly "universal" solvent. It is a polar liquid capable of dissolving most ionic and polar covalent compounds—and even some nonpolar covalent compounds to some extent! Because of the structure of ionic compounds, water breaks them down into their component ions (unlike covalent compounds, which do not usually break down when dissolved in water). For example, sodium chloride breaks down into ions, or **dissociates**, when dissolved (Eq. 13):

$$NaCl(s) \xrightarrow{\;H_2O\;} Na^+(aq) + Cl^-(aq) \qquad [13]$$

Thus, when an ionic compound is dissolved in water, it undergoes **dissociation** and actually exists in the form of separated ions (like the $Na^+(aq)$ and $Cl^-(aq)$ in Equation 13, not in the combined form of $NaCl(aq)$). However, not all ionic compounds dissolve equally well in water. The difference is in how strong the attraction forces between water and the ions are compared to the strength of the ionic bond. If the attraction to water is stronger than the hold of the ionic bond, most of the ionic compound will dissociate and exist as aqueous ions (like NaCl). We call these ionic compounds **soluble** in water.

On the other hand, if the ionic bond is stronger than the attraction between ions and water, most of the compound remains intact, with only a few ions separating and floating around in water. We call these types of ionic compounds **insoluble**. In fact, when we put ions of insoluble compounds together in one solution, they will combine and form the solid ionic compound. When a solid is formed from the addition of two solutions of ionic compounds because an insoluble product is formed, we say that the ionic compound **precipitated** (fell down to the bottom) from the solution. The process of two ions dissolved in water coming together to form a solid, shown in Equation 14 below, is called **precipitation**, and the solid itself is called a **precipitate**:

$$Pb^{2+}(aq) + 2\ Cl^-(aq) \rightarrow PbCl_2(s) \qquad [14]$$

If we look back at Eq. 12, we will see that $PbCl_2$ is one of the products of that reaction—and is, in fact, the only product that is a solid. We would call Eq. 12 a precipitation reaction as well, since a precipitate is formed! Now that we know that soluble ionic compounds exist in the form of ions, we could actually rewrite Eq. 12, replacing any soluble ionic compound with the (aq) designation with its ions (as Equation 15 shows):

$$Pb(NO_3)_2(aq) \quad + \quad 2\ KCl(aq) \quad \rightarrow PbCl_2(s)\ + \quad 2\ KNO_3(aq)$$

$$Pb^{2+}(aq) + 2\ NO_3^-(aq) + 2\ K^+(aq) + 2\ Cl^-(aq) \rightarrow PbCl_2(s) + 2\ K^+(aq) + 2\ NO_3^-(aq) \qquad [15]$$

The chemical equation shown in Eq. 15 describes the same double displacement reaction as Eq. 12—but in a different way. Eq. 12 shows the chemical formulas of each of the reactants dissolved in water and provides an "overview" or "big picture" of the chemical reaction: a solution of $Pb(NO_3)_2$ is added to a solution of KCl, and as a result the solid $PbCl_2$ and a solution of KNO_3 above it are produced. Since all the chemical formulas are written "together" and not broken up into ions, like Eq. 15, this way of writing the equation is called a **"molecular" equation**. (The word "molecular" is used loosely, since there aren't any actual *molecules* in ionic compounds!)

On the other hand, Eq. 15 shows *what is actually happening in solution*—we start out with two pairs of dissociated ions, and two of these react to form the solid precipitate, while the other two just "stay there." This type of equation, showing the separated ions for aqueous ionic compounds, is called the **complete ionic equation**.

We can describe this reaction in one more way: the ions that do not get attracted to each other do not really change and therefore do not take any part in the reaction. We called these ions **spectator ions** and can simply omit them:

$$Pb^{2+}(aq) + \cancel{2\ NO_3^-(aq)} + \cancel{2\ K^+(aq)} + 2\ Cl^-(aq) \rightarrow PbCl_2(s) + \cancel{2\ K^+(aq)} + \cancel{2\ NO_3^-(aq)}$$

$$Pb^{2+}(aq) \qquad + \qquad 2\ Cl^-(aq) \rightarrow PbCl_2(s)$$

The resulting equation is an exact replica of Eq. 14! This type of equation, showing only the ions that take an active part in the reaction and omitting the spectator ions, is called the **net ionic equation**. Thus, we can see that the chemical equations for double displacement reactions can be written in three ways:

- The "molecular" equation, with all ionic compounds written "together" as formulas;
- The complete ionic equation, with any (aq) components rewritten as separate ions;
- The net ionic equation, similar to complete ionic but omitting the spectator ions.

As mentioned above, precipitation reactions occur because an insoluble product is formed. If both products were soluble in water, *all ions* would be spectators, resulting in no reaction (Eq. 16-17):

$$\text{NaNO}_3(\text{aq}) \quad + \quad \text{KCl}(\text{aq}) \quad \rightarrow \quad \text{NaCl}(\text{aq}) \quad + \quad \text{KNO}_3(\text{aq}) \qquad [16]$$

$$\text{Na}^+(\text{aq}) + \cancel{\text{NO}_3^-(\text{aq})} + \cancel{\text{K}^+(\text{aq})} + \text{Cl}^-(\text{aq}) \rightarrow \text{Na}^+(\text{aq}) + \text{Cl}^-(\text{aq}) + \cancel{\text{K}^+(\text{aq})} + \cancel{\text{NO}_3^-(\text{aq})} \qquad [17]$$

In general, there are three options for what can happen to cause a double displacement reaction to happen:

1. One of the products is a precipitate (as shown above)
2. One of the products is a gas
3. One of the products is water—$H_2O(\ell)$

We can also write ionic equations for single displacement reactions, since they, too, have aqueous solutions of ionic compounds! For example, Eq. 8 could be rewritten this way as the complete ionic (Eq. 18) and net ionic (Eq. 19) equations:

$$2\,\text{Al}(\text{s}) \quad + \quad 6\,\text{HNO}_3(\text{aq}) \quad \rightarrow \quad 2\,\text{Al}(\text{NO}_3)_3(\text{aq}) \quad + \quad 3\,\text{H}_2(\text{g})$$

$$2\,\text{Al}(\text{s}) + 6\,\text{H}^+(\text{aq}) + \cancel{6\,\text{NO}_3^-(\text{aq})} \rightarrow 2\,\text{Al}^{3+}(\text{aq}) + \cancel{6\,\text{NO}_3^-(\text{aq})} + 3\,\text{H}_2(\text{g}) \qquad [18]$$

$$2\,\text{Al}(\text{s}) + 6\,\text{H}^+(\text{aq}) \quad \rightarrow \quad 2\,\text{Al}^{3+}(\text{aq}) + \quad 3\,\text{H}_2(\text{g}) \qquad [19]$$

In the write-up part of this experiment, you will actually practice writing the "molecular" and ionic equations for some of the reactions you will see demonstrated by the instructor during this experiment—as well as some additional ones.

Solubility of Ionic Compounds

As discussed above, for double displacement reactions it is important to know whether a product will be **insoluble** in water, thus forming a **precipitate**. It would be fairly difficult to *predict* whether the attraction between a specific cation and anion would be stronger (resulting in a precipitate) or weaker (resulting in an aqueous solution of the ionic compound). Therefore, chemists use tables of experimental data, showing how soluble each particular compound is. There are several ways to show this numerically—for example, by expressing solubility in grams of solute per 100 g H_2O or as molarity, etc. These data can also be justified theoretically, and some trends can be identified and memorized.

However, for us, it would be enough to "roughly" separate the "**soluble**" compounds, which easily and readily dissolve in water, and "**insoluble**" ones, which <u>do</u> dissolve in water but only to a very small extent. Of course, this is an oversimplification—but one that will work well for our purposes! These data are summarized in Table 1. Please note the legend for the symbols listed below the table itself! Of most interest to you will be: "**aq**," meaning the compound is soluble and exists in water as dissociated aqueous ions; and "**s**," meaning the compound is insoluble and does not dissolve in water very much—so that when two solutions are combined, each of which contains one of the two ions, the insoluble compound would precipitate.

Table 1. Solubility chart

	Br^-	CH_3COO^-	CO_3^{2-}	Cl^-	ClO_3^-	CrO_4^{2-}	F^-	I^-	NO_3^-	O^{2-}	OH^-	PO_4^{3-}	SO_4^{2-}	S^{2-}
Ag^+	s	aq	s	s	aq	s	aq	s	aq	s	-	s	s	s
Al^{3+}	aq	aq	-	aq	aq	aq	aq	aq	aq	s	s	s	-	D
Ba^{2+}	aq	aq	s	aq	aq	s	s	aq	aq	aq	aq	s	s	aq
Ca^{2+}	aq	aq	s	aq	aq	aq	s	aq	aq	s	**aq/s**	s	s	s
Cd^{2+}	aq	aq	s	aq	aq	aq	aq	aq	aq	s	s	s	aq	s
Cu^+	aq/s	-	s	s	-	aq		s	-	s	s	-	D	s
Cu^{2+}	aq	aq	D	aq	aq	s	s	-	aq	s	s	s	aq	s
Fe^{2+}	aq	aq	s	aq	-	-	s	aq	aq	s	s	s	aq	s
Fe^{3+}	aq	s	-	aq	-	s		-	aq	s	s	s	aq/s	s
H^+	**aq**	aq	aq	**aq**	**aq**	aq	aq	**aq**	**aq**	aq	H_2O	aq	**aq**	aq
Hg^+	s	aq/s	s	s	aq/s	s	-	aq/s	aq/D	s	-	s/D	s	s
Hg^{2+}	aq/s	aq	s	aq	aq	aq		s	aq	s	s	aq/s	D	s
K^+	aq	aq	aq	aq	aq	aq	aq	aq	aq	D	**aq**	aq	aq	s
Mg^{2+}	aq	aq	s	aq	aq	aq	s	aq	aq	s	s	s	aq	D
Mn^{2+}	aq	aq	s	aq	-	aq	s	aq	aq	s	s	-	aq	s
Na^+	aq	aq/s	aq	aq	aq	aq	aq	aq	aq	D	**aq**	aq	aq	s
NH_4^+	aq	aq	aq	aq	aq	aq	aq	aq	aq	-	aq	aq	aq	aq
Ni^{2+}	aq	aq	s	aq	aq	aq	s	aq	aq	s	s	s	aq	s
Pb^{2+}	aq/s	aq	s	aq/s	aq	s	s	s	aq	s	s	s	s	s
Pb^{4+}	-	D	-	D	-	-	-	-	-	s	-	-	-	-
Zn^{2+}	aq	aq	s	aq	-	aq	s	aq	aq	s	s	s	aq	s

Key: 'aq' soluble   's' insoluble   '-' does not exist   'aq/s' partly soluble   'D' decomposes

Objectives

1. To observe demonstrations of combination, decomposition, single displacement, and double displacement reactions

2. To balance chemical equations

3. To categorize chemical reactions as combination, decomposition, single displacement, or double displacement

4. To complete chemical equations for combination, single displacement, and double displacement reactions

5. To write complete ionic and net ionic equations for single displacement and double displacement reactions

6. To complete chemical equations for observed reactions

Safety

- Wear approved **safety glasses or goggles** during the entire experiment!

- Avoid **burns** by contact with hot objects (hot plate, boiling water)! Never hand hot objects to another person!

- Many of the chemicals used in the demonstration and experiment are potentially harmful! Prevent contact with your eyes, skin, or clothing. Avoid swallowing any of the reagents.

- Wash your hands thoroughly with soap before you leave the lab room!

Materials and Equipment

You will need the following for your experiment:

1. Magnesium pieces
2. Hydrochloric acid, 3 M solution
3. Lead(II) nitrate solution
4. Potassium iodide solution
5. Hydrochloric acid, 0.1 M solution
6. Sodium carbonate solution
7. Sodium hydroxide, 0.1 M solution
8. Phenolphthalein
9. Steel wool
10. Copper(II) sulfate, 0.1 M solution
11. Copper(II) sulfate pentahydrate
12. Ammonium carbonate
13. Litmus paper, red
14. Test tubes (7)
15. Glass stirring rod
16. Plastic pipette
17. Test tube tongs

Procedure

Record the appearance of the substances before each reaction, as well as the change(s) observed during each. Pay special attention to the following: gas formation (bubbles), solid precipitate formation (cloudiness), color changes, and temperature changes.

1. Use a scoopula to place a piece of magnesium metal (Mg) into a test tube. Then place a pipetteful of hydrochloric acid (3 M HCl) into the same test tube. Observe and record any changes.

2. Place one pipetteful of potassium iodide (KI) into a test tube. Then put 1-2 drops of lead(II) nitrate ($Pb(NO_3)_2$) into the same test tube. Observe and record any changes.

3. Place one pipetteful of sodium carbonate (Na_2CO_3) into a test tube. Then place one pipetteful of hydrochloric acid (0.1 M HCl) into the same test tube. Observe and record any changes.

   Repeat Step 3 once more: Place one pipetteful of sodium carbonate (Na_2CO_3) into a test tube. This time, place 1-2 drops of phenolphthalein into the test tube after the Na_2CO_3 has been added (but before the HCl!). Then place one pipetteful of hydrochloric acid (0.1 M HCl) into the same test tube. Once again, observe and record any changes.

   *Phenolphthalein is a so-called **acid-base indicator**, and it changes its color to **pink** when the solution is basic (alkaline) or to **colorless** when it is acidic.*

4. Place one pipetteful of sodium hydroxide (NaOH) into a test tube. Then place one pipetteful of hydrochloric acid (0.1 M HCl) into the same test tube. Observe and record any changes.

    Repeat Step 4 once more: Place one pipetteful of sodium hydroxide (NaOH) into a test tube. This time, place 1-2 drops of phenolphthalein into the test tube after the NaOH has been added (but before the HCl!). Then place one pipetteful of hydrochloric acid (0.1 M HCl) into the same test tube. Once again, observe and record any changes.

    See the note under Step 3 above for an explanation of the colors.

5. Cut a small piece of steel wool (iron, Fe) and roll it up into a ball. Place it inside a test tube and then place one pipetteful of copper(II) sulfate ($CuSO_4$) into the same test tube. Shake gently to stir for 1-3 minutes, observing and recording any changes.

6. Place a small amount of copper(II) sulfate pentahydrate ($CuSO_4 \cdot 5\ H_2O$) on the end of a spatula and pour it inside a ***clean, dry*** test tube. Using your test tube tongs, hold the test tube in the flame of a Bunsen burner to heat the solid. Observe and record any changes. Pay special attention to what might appear on the sides of the test tube. (Once you are done with the heating and the test tube has cooled, you may want to add a few drops of water to the solid in the test tube to see another change!)

7. Place a small amount of ammonium carbonate (($NH_4)_2CO_3$) on the end of a spatula and pour it inside a **clean, dry** test tube. Using your test tube tongs, hold the test tube in the flame of a Bunsen burner to heat the solid. Observe and record any changes. Pay special attention to any odor and to what might appear on the sides of the test tube. Also, take a piece of red litmus paper, wet it slightly, and place it at the mouth of the test tube while it is being heated; note any changes in the litmus paper.

    *The litmus paper is another **acid-base indicator**, changing color to **pink/red** in the presence of an acid or to **blue** in the presence of a base.*

Worksheet Directions

Reactions from the Experiment

Based on the observations you made in the experiment, write balanced equations for the reactions you observed. Reactions are numbered the same way they are on your Data Sheet.

Include complete and net ionic equations for single and double displacement reactions. If you need more instructions on how to do this, you can refer back to the Background Section or this lab or Appendix II at the back of the lab book.

Eq 1. This is a ***single displacement***, where hydrogen and magnesium switch. Write correct chemical formulas for the products (use the octet rule and the "X rule"): Since hydrogen is a ***gas***, this explains the bubbles you observed—and hydrogen is a ***diatomic element*** (see Eq. 20 for an additional hint!). In strong enough acid, magnesium *metal* completely disappears and forms a compound with the Cl^- ion. This compound is ***soluble*** in water!

   Also give the complete ionic and net ionic equations for this reaction!

Eq 2. Lead(II) nitrate and potassium iodide react in this ***double displacement*** reaction. Write the correct chemical formulas for the products (use the octet rule and the "X rule"). Use the Solubility Chart (Table 1) to determine which product is a ***precipitate*** (a solid!).

   Also give the complete ionic and net ionic equations for this reaction!

Eq 3. This reaction is a bit different! Although you can write a ***double displacement*** equation with aqueous sodium chloride and $H_2CO_3(aq)$ as products—which should indicate no reaction—you *know* there was a reaction because you observed bubbles! $H_2CO_3(aq)$ actually ***decomposes*** and forms CO_2 gas and liquid H_2O. Therefore, the three products of this reaction are: aqueous sodium chloride, gaseous carbon dioxide, and liquid water.

Also give the complete ionic and net ionic equations for this reaction!

Eq 4. This reaction is actually a quite simple ***double displacement*** reaction. Write the correct chemical formulas for the products (use the octet rule and the "X rule")—just remember what happens when the hydrogen ion combines with OH^-!

Also give the complete ionic and net ionic equations for this reaction!

Eq 5. This is another ***single displacement*** reaction, where Cu and Fe change places. Assume that iron forms a Fe^{2+} ion (as opposed to Fe^{3+}) and write the correct chemical formulas for the products (use the octet rule and the "X rule"). The elemental copper metal is solid, while the ionic compound of iron(II) ions and sulfate ions is soluble (see Table 1).

Also give the complete ionic and net ionic equations for this reaction!

Eq 6. This is a ***decomposition*** reaction—recall that you observed water <u>condensing</u> on the sides of the test tube (which means water was produces in its gaseous form, as "steam" or vapor), and the remaining white solid is the same ionic compound—minus the 5 H_2O!

*Since none of the products are aqueous solutions of ionic compounds, you **<u>cannot write</u>** complete ionic and net ionic equations for this reaction!*

Eq 7. This is another ***decomposition*** reaction—recall that you observed water <u>condensing</u> on the sides of the test tube (which, again, means water was produced in its gaseous form, as "steam" or vapor) and smelled the ammonia gas (NH_3) that was formed. The third product of this reaction, which we could not really observe, is carbon dioxide gas.

*Since none of the products are aqueous solutions of ionic compounds, you **<u>cannot write</u>** complete ionic and net ionic equations for this reaction!*

____   __________   __________
name   section   date

Pre-Laboratory Assignment

1. Determine the reaction type for the following reactions:

                                                              Reaction Type

    a. $2\ Al(s) + 3\ H_2SO_4(aq) \rightarrow Al_2(SO_4)_3(aq) + 3\ H_2(g)$ _____________________

    b. $CaCO_3(s) \rightarrow CaO(s) + CO_2(g)$ _____________________

    c. $Pb(NO_3)_2(aq) + K_2CrO_4(aq) \rightarrow PbCrO_4(s) + 2\ KNO_3(aq)$ _____________________

    d. $4\ Fe(s) + 3\ O_2(g) \rightarrow 2\ Fe_2O_3(s)$ _____________________

    e. $Mg(OH)_2(aq) + H_3PO_4(aq) \rightarrow Mg_3PO_4(aq) + HOH(\ell)$ _____________________

2. Define the following terms, as they apply to this experiment:

    a.   Precipitate –

    b.   Complete ionic equation –

    c.   Spectator ion –

    d.   Dissociation –

name   section   date

Data/Observation Sheet

Reactants	Appearance Before	Change(s) Observed
1. $Mg(s) + HCl(aq) \rightarrow$ ...	Mg: HCl:	
2. $Pb(NO_3)_2(aq) + KI(aq) \rightarrow$ ...	$Pb(NO_3)_2$: KI:	
3. $HCl(aq) + Na_2CO_3(aq) \rightarrow$ ...	HCl: Na_2CO_3:	Repeat with phenolphthalein:
4. $HCl(aq) + NaOH(aq) \rightarrow$ ...	IICl: NaOH:	Repeat with phenolphthalein:
5. $CuSO_4(aq) + Fe(s) \rightarrow$ ...	$CuSO_4$: Fe:	
6. $CuSO_4 \cdot 5\,H_2O(s) \xrightarrow{\Delta}$ ...		
7. $(NH_4)_2CO_3(s) \xrightarrow{\Delta}$ ...		

name   section   date

Worksheet

Reactions from the Experiment

Complete and balance the chemical equations for the observed reactions. Where necessary, write the complete ionic and net ionic equations (for single and double displacement). Details of the reaction including reactants, products and the type of reaction it is are given in the Worksheet Directions section.

Eq. 1. _____ $Mg(s)$ + _____ $HCl(aq)$ $\rightarrow$

Eq. 2. _____ $Pb(NO_3)_2(aq)$ + _____ $KI(aq)$ $\rightarrow$

Eq. 3. _____ HCl(aq) + _____ Na$_2$CO$_3$(aq) $\rightarrow$

Eq. 4. _____ HCl(aq) + _____ NaOH(aq) $\rightarrow$

Eq. 5. _____ $CuSO_4(aq)$ + _____ $Fe(s)$ →

Eq. 6. _____ $CuSO_4 \cdot 5\ H_2O(s)$ $\xrightarrow{\Delta}$

Eq. 7. _____ $(NH_4)_2CO_3(s)$ $\xrightarrow{\Delta}$

EXPERIMENT 6: **Mass Relationships in Reactions**

Mike Goldin (Liberty University)

To everything there is a season, a time for every purpose under heaven…
A time to break down, and a time to build up.
– Ecclesiastes 3:1,3b (NKJV)

Background

A **chemical reaction** is a process of breaking bonds in the **reactants** (starting substances) and making new bonds between the atoms, resulting in the formation of new substances called **products**. In Experiment 4, we described the process of burning natural gas (methane, CH_4) by reacting it with oxygen gas (O_2) and producing two new substances, carbon dioxide (CO_2) and water (H_2O), by writing a balanced chemical equation for this reaction (Eq. 1):

$$CH_4 + 2\,O_2 \rightarrow CO_2 + 2\,H_2O \qquad\qquad [1]$$

This equation represents what happens at the molecular level, using the chemical formulas of the reactants and products. The equation shows that for each molecule of CH_4 has to react with two molecules of O_2 to produce one molecule of CO_2 and two molecules of H_2O. Similarly to mathematical equations, the numbers showing how many molecules are needed are called **coefficients**, since they multiply each of the chemical formulas. However, in lab, we must have a way of tying these chemical equations with real quantities we would see in lab! For example, we know that one molecule of CH_4 in Equation 1 would produce one molecule of CO_2 and two molecules of H_2O when reacted with oxygen; but if we started with 2.50 grams of CH_4, how many grams of CO_2 or H_2O would we get? There are a couple of issues to consider here: first of all, each molecule will have a different mass since each type of atom has a different atomic weight and secondly, the $CH_4 : O_2 : CO_2 : H_2O$ ratios of 1:2:1:2 are ratios of molecule counts, ***not*** masses!

Masses of individual atoms in the Periodic Table are measured in atomic mass units (amu). Masses of molecules in amus can be easily calculated: for example, a molecule of CH_4 has a mass of 12.011 amu + 4(1.008 amu) = 16.043 amu. The masses in amus can then be converted to grams, but 1 amu is equal to 1.66×10^{-24} grams—an extremely small number! It is much more useful and practical to somehow relate this to *measurable* masses and to the relationships in the number of molecules. This is why chemists have come up with a unit to count particles called the **mole**—a quantity equal to 6.02×10^{23} particles. For example, for CH_4 molecules:

$$1 \text{ mole } CH_4 = 6.02 \times 10^{23} \text{ molecules } CH_4 \qquad\qquad [2]$$

The number 6.02×10^{23} (called **Avogadro's number**) might seem random, but it is actually quite the opposite! The way this number is chosen allows us to convert the mass of *one molecule* (in amu) to the mass of *one mole* (in grams) by simply taking the number in amu and changing the units to grams! This is because $(6.02 \times 10^{23}) \times (1.66 \times 10^{-24}) = 1$. In other words, for CH_4:

$$1 \text{ molecule } CH_4 = 16.043 \text{ amu } CH_4 \qquad\qquad [3a]$$

$$1 \text{ mole } CH_4 = 16.043 \text{ g } CH_4 \qquad\qquad [3b]$$

This makes it extremely easy to calculate the mass of 1 mole of the substance, called the **molar mass** of the substance and measured in grams per mole (g/mol). Thus, the molar mass of CH_4 is equal to 16.043 g/mol. This number provides the relationship needed between the number of molecules (measured as moles) and masses in grams. Now the number of grams of H_2O produced from 2.50 g of CH_4, because we know that the ratio of CH_4 can be calculated: H_2O = 1:2,

or:

$$1 \text{ molecule } CH_4 = 2 \text{ molecules } H_2O \qquad \textbf{[4a]}$$

$$1 \text{ mol } CH_4 = 2 \text{ mol } H_2O \qquad \textbf{[4b]}$$

1st step: Convert 2.50 g of CH_4 to moles:

$$2.50 \text{ g } CH_4 \times \frac{1 \text{ mol } CH_4}{16.043 \text{ g } CH_4} \ \ldots \qquad \textbf{[5a]}$$

then convert the moles of CH_4 to moles of H_2O:

$$2.50 \text{ g } CH_4 \times \frac{1 \text{ mol } CH_4}{16.043 \text{ g } CH_4} \times \frac{2 \text{ mol } H_2O}{1 \text{ mol } CH_4} \ \ldots \qquad \textbf{[5b]}$$

Finally, convert the moles of H_2O to grams of H_2O—which is what we are looking for!

$$2.50 \text{ g } CH_4 \times \frac{1 \text{ mol } CH_4}{16.043 \text{ g } CH_4} \times \frac{2 \text{ mol } H_2O}{1 \text{ mol } CH_4} \times \frac{18.015 \text{ g } H_2O}{1 \text{ mol } H_2O} = 5.61 \text{ g } H_2O \qquad \textbf{[5c]}$$

(The molar mass of water is 2(1.008)+15.999 = 18.015 g/mol, based on atomic weights.)

This calculated value of expected product amount for the reaction in Equation 1 is called the **theoretical yield** of H_2O—how much water will be produced if all of the 2.50 grams of CH_4 would react with oxygen completely. In reality, we rarely can get 100% of any reactant to fully convert to products. The actual amount of water produced from this reaction is called the **actual yield**, and will likely be be smaller than the theoretical.

In this experiment, magnesium metal (in the form of either pieces or strip) will be heated it in the air (in a crucible) to allow it to react with oxygen according to this reaction (Eq. 6):

$$2 \text{ Mg} + O_2 \ \rightarrow \ 2 \text{ MgO} \qquad \textbf{[6]}$$

The mass of magnesium metal (Mg) will be measured before the reaction starts. It will then be heated until all of it (theoretically) is converted to magnesium oxide (MgO). This product will be weighed. The mass of the original Mg will be used along with the balanced chemical equation to calculate the theoretical yield of MgO, as shown in Equation 5(a-c).This theoretical yield value will then be compared to the measured actual yield by calculating the percent yield (Eq. 7):

$$\% \text{ yield} = \frac{\text{actual yield (g)}}{\text{theoretical yield (g)}} \times 100\% \qquad \textbf{[7]}$$

Objectives

1. To prepare a metal oxide
2. To determine the percent yield of a chemical reaction
3. To determine the experimental chemical formula (empirical formula) of a compound

Safety

*   Wear approved **safety glasses or goggles** during the entire experiment!

*   Avoid **burns** by contact with hot objects (open flame, crucibles and lids, etc.)! NEVER hand hot objects to another person!

*   Magnesium can "flash" while it is being heated in the crucible—**do not look at the flash!** Immediately cover up the crucible completely with the lid!

*   **Crucibles break easily**, since they are made of porcelain! Hold crucibles and lids over an empty beaker when carrying them. If one breaks, please use the broom and dustpan for broken glass to pick up the pieces; dispose of them in the broken glass container.

- **Never place a hot crucible on the benchtop!** Always use a hot pad or leave it in the clay triangle to cool.

Materials and Equipment

You will need the following for your experiment:

1. Magnesium ribbon (or pieces)
2. Crucible and cover
3. Crucible tongs (or hot mitt)
4. Bunsen burner
5. Rubber or plastic hose
6. Ring support clamp
7. Clay triangle
8. Striker
9. Glass stirring rod

Procedure

1. Get all of the equipment listed above in preparation for the experiment.

2. Clamp the ring support on the metal rod at your lab bench. Place the clay triangle on top of it, with the Bunsen burner directly underneath it. Connect the Bunsen burner to the gas valve on the lab bench with the rubber (or plastic) hose.

3. Because of the tendency of the porcelain that the crucibles are made of to absorb moisture from the air, the crucible and lid have to be heated before the masses are obtained. To do this, place the crucible on the clay triangle, as shown in Figure 1. Place the lid on top of it slightly ajar, so the crucible is not completely covered. Light the Bunsen burner and adjust the height and position of the ring support, so that the tip of the flame barely touches the bottom of the crucible; heat the crucible for 2-3 minutes.

Figure 1. Crucible setup

4. After 2 minutes, adjust (increase) the flame, so the top of the hottest *inner blue cone* of the flame is touching the bottom of the crucible. Continue heating like this for another 5-7 minutes.

5. Allow the crucible to cool for 10 minutes until it is no longer hot to the touch. Carefully use crucible tongs to cover the crucible completely with the lid and lift the crucible out of the clay triangle by using the tongs.

6. Weigh the crucible and cover; record the mass (with proper sig. figures) on the Data Sheet.

> *IMPORTANT: Do not weigh a hot crucible on the balance or place it on the lab bench. This will result in damage to both, as well as inaccurate weighing. The crucible must be cool enough to hold in your hand, so remember to "hover" your hand over it and, if you don't feel the heat, gently and carefully touch it*

7. Cut three 1-inch long pieces of magnesium ribbon (or take 6-7 pieces of magnesium turnings) and add to the crucible. Weigh the metal inside the crucible with the cover on; record the mass with proper significant figures on the Data Sheet.

8. Place the crucible with the metal and cover back onto the clay triangle and set the lid slightly ajar. Heat the crucible for 10 min in the hottest flame, with the inner blue cone of the flame touching the bottom of the crucible (which should start glowing red).

> **CAUTION! The magnesium will sometimes react very quickly with the oxygen in the air before the 10 minutes are up, causing it to "flash" and give off bright light. If this happens, _DO NOT LOOK AT THE LIGHT_! Immediately cover the crucible completely with the lid and turn down the flame. After at least 30 seconds, move on to Step 9.**

9. Carefully lift the cover off the crucible with the tongs and check the metal after 10 minutes. Try breaking up the solid using a glass stirring rod: if any large pieces are still there that will not break up into powder, continue heating until all of the metal has changed into the dull-colored, grayish powder-like magnesium oxide.

10. Turn off the burner and allow the crucible to cool for 10 minutes until it is no longer hot to the touch. Carefully use crucible tongs to cover the crucible completely with the lid and lift the crucible out of the clay triangle by using the tongs.

11. Weigh the crucible and cover with the magnesium oxide and record the mass (with proper significant figures) on the Data Sheet.

> **IMPORTANT: Do not weigh a hot crucible on the balance or place it directly on the lab bench. This will result in damage to both, as well as inaccurate weighing.**

Clean-Up

- Discard any solid from the crucible in the designated collection container.

- Wash the crucible and cover and dry them with a paper towel.

- Clean up any spilled powder off your lab bench with a wet paper towel. Then dry the benchtop with a dry paper towel.

- Return the following items to the front desk:
  - crucible;
  - crucible cover;
  - Bunsen burner;
  - rubber hose;
  - clay triangle;
  - striker;
  - crucible tongs;
  - hot mitt.

Calculations

Use the data from your Data Sheet to do these calculations. Record results in the Calculations section of the Data Sheet.

1. Calculate the mass of magnesium in your initial sample (magnesium strip or pieces) by difference, using the mass of the empty crucible and cover.

2. Calculate the mass of magnesium oxide at the end of the experiment by difference, using the mass of the empty crucible and cover.

3. Calculate the theoretical yield of magnesium oxide based on the initial mass of magnesium, its molar mass, the ratio of coefficients (Mg : MgO), and the molar mass of MgO (see Eq. 5 for a worked example for another reaction). Use the coefficients from Eq. 6.

4. Calculate the percent yield of magnesium oxide (according to Eq. 7).

Pre-Laboratory Assignment

1. Define the following terms, as they apply to this experiment:

    a.   Chemical reaction –

    b.   Mole –

    c.   Reaction coefficient –

    d.   Molar mass –

    e.   Theoretical yield –

    f.   Actual yield –

(over, please)

2. 2.56 g of aluminum foil, Al, reacts with unlimited supply of oxygen gas, O_2, in the air to produce aluminum oxide, Al_2O_3:

$$4\ Al + 3\ O_2\ \rightarrow\ 2\ Al_2O_3$$

    a.   What is the molar mass of aluminum, Al, and aluminum oxide, Al_2O_3? *Show your work and follow the rules to obtain the correct number of significant figures in the result.*

Molar mass of Al: _________________

Molar mass of Al_2O_3: _________________

    b.   From the coefficients in the chemical equation above, fill in the numbers in the conversion factor below:

$$\frac{\text{moles } Al_2O_3}{\text{moles } Al}$$

    c.   Follow the pattern in Equation 5c (on p. 130) to calculate how many grams of Al_2O_3 would be formed from the 2.56 g of Al. Use the numbers you have written down in parts a and b to help you fill in the conversions. This is the **theoretical yield** of Al_2O_3! *Show your work and follow the rules to obtain the correct number of significant figures in the result.*

name   section   date

Data Sheet

1. Mass of empty crucible + cover (after pre-heating), g   ___________________

2. Mass of crucible + cover + magnesium, g   ___________________

3. Mass of crucible + cover + magnesium oxide, g   ___________________
   (This value is obtained at the very end of the experiment)

Calculations

Mass of magnesium (Mg), g   ___________________
   Show work:

Mass of magnesium oxide (MgO), g   ___________________
   Show work:

Theoretical yield of MgO, g   ___________________
   Show work:

Percent yield of MgO, %   ___________________
   Show work:

_______________________________   ________   _______
name   section   date

Post-Laboratory Assignment

The experiment we performed can also be used to determine the experimental formula of magnesium oxide. In this post-lab assignment, you will take the data and calculations from your experiment, do a few more calculations, and determine the formula of magnesium oxide based on the experimentally determined ratio of moles Mg : moles O *in this experiment.*

1. Calculate the mass of oxygen **atoms** in magnesium oxide by subtracting the original mass of magnesium (line 1 on Data Sheet) from the mass of magnesium oxide formed (line 3 on Data Sheet). *Show work!*

    Mass of Mg, g: ________________   Mass of MgO, g: ________________

    Mass of O, g: ________________

2. Use the mass of oxygen atoms determined in number 1 above and the mass of Mg measured at the beginning of the lab to calculate the number of moles of Mg atoms and O **atoms** in magnesium oxide based on their masses and molar masses. Use the molar mass of Mg and the molar mass of O to calculate the number of moles of Mg atoms and O **atoms** in magnesium oxide based on their masses and molar masses. Remember that the molar mass of each is used to convert from mass to moles. *Show work!*

    Mass of Mg, g: ________________   Molar mass of Mg, mol/g: ________________

    Number of moles Mg: ________________

    Number of moles O: ________________

3. Find the ratio of moles Mg : moles O by dividing both amounts calculated in Step 2 by the smaller of the two numbers. Round them to the nearest whole number. *Show work!*

    Whole-number ratio of Mg:O ___________ : ___________

4. Based on the *calculated* whole-number ratio from Step 3, write the chemical formula for magnesium oxide. *HINT:* A chemical formula is made up of the symbols of each type of atom (Mg and O) and subscripts that indicate the whole–number ratio of the types of atoms.

    Experimental chemical formula of magnesium oxide: _______________________

5. The true formula of magnesium oxide is MgO. Was your experimentally obtained chemical formula the same as the true formula? If not, discuss what experimental problems may have occurred to cause this discrepancy. Think this through on your own, based on what was done in the experiment.

Mike Goldin (Liberty University)

I go forward, but He is not there, and backward, but I cannot perceive Him
– Job 23:8 (NKJV)

Background

From Experiment 4, you should have a fairly good grasp of what happens with ionic compounds when they are dissolved in water. This will help tackle the central part of this experiment—the concept of reversible chemical reactions (and processes). A **reversible** process is a reaction or physical change that can go forward or backward. For example, both the melting/freezing of water (Eq. 1) and the dissolution/precipitation of lead(II) chloride (Eq. 1) are reversible processes:

$$H_2O(s) \rightleftharpoons H_2O(\ell) \qquad\qquad\qquad [1]$$

$$PbCl_2(s) \rightleftharpoons Pb^{2+}(aq) + 2\ Cl^-(aq) \qquad\qquad [2]$$

Notice that the equations here are written with a double arrow "$\rightleftharpoons$". This symbol indicates that indicates the process is reversible! If the equation is read from left to right, the process described by them is the **forward reaction** (or process); for $PbCl_2$ it would be:

$$PbCl_2(s) \rightarrow Pb^{2+}(aq) + 2\ Cl^-(aq)$$

If the equation is read from right to left, however, the **reverse reaction** (or process) is occurring:

$$Pb^{2+}(aq) + 2\ Cl^-(aq) \rightarrow PbCl_2(s)$$

In order to really grasp what occurs to cause these processes to be reversible, the reaction with $PbCl_2$ will be further examined. Imagine that a few crystals of solid $PbCl_2$ is added to a beaker that contains pure water in it. At time zero, when water has not yet started to interact with the solid and pull some of the ions away, so only water and the solid are present (Fig. 1a). After some time, water molecules are able to come in and begin breaking a small amount of the solid into ions (Fig. 1b). As additional time passes, more ions appear and more solid disappears (Fig. 1c). We will now pay special attention to the _speed_ (or _rate_) of this process!

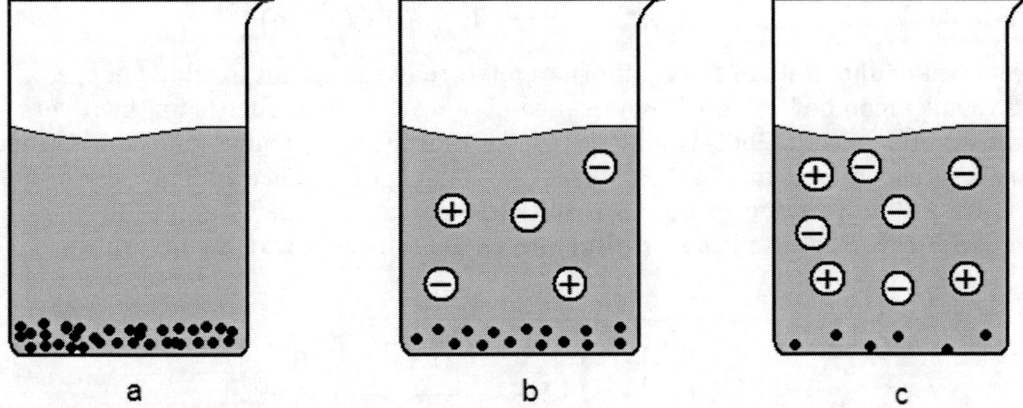

Figure 1. Dissolution of PbCl$_2$ in the process of time

(Of course, we must note that the pictures in Fig. 1 are not completely accurate—the Pb^{2+} ion has a 2+ charge, and the ions and particles of solid are drawn way bigger than they really are! This is simply a "cartoon" or a conceptual picture of what is happening.)

As more and more ions dissolve in water, they have more and more chances to collide with each other. When this happens, they can get attracted to one another and "stick together" as they reestablish their ionic bond—which is the reverse process. As more and more ions from the solid continuing to dissolve, the likelihood of them recombining and precipitating as solid increases! We could say that *the reverse process is speeding up* with passage of time, whereas the forward process continues at the same speed regardless of how much solid there is.

This is true because the rate of dissolution of a solid usually only depends on temperature. (Frequently, however, *the forward process is slowing down* in speed with passage of time when it does not involve a solid dissolving.)

Since the reverse process speeds up in time, there will come a time when the two speeds will "catch up" with each other. At that point, however many ions are continually produced by the forward process will also recombine and precipitate in the reverse process. The ratio of the amount of solid to the amount of ions at that point will stop changing—since the two processes essentially "cancel out!" This condition, when the rates (or speeds) of the forward and reverse processes are equal is called **equilibrium** (also **chemical equilibrium** or **dynamic equilibrium**).

Note that neither the forward nor the reverse reaction ever stop—they simply continue on (in theory, forever!) at equal rates, and therefore there is no more change in the amounts of reactants or products, since their ratio (proportion) gets "stuck" at a certain value. Therefore, the reaction *will never have 100% yield*, since it will always contain some **mixture of the reactants and products**! This is a very important point about reversible reactions. Only if the products are continually removed as they are made will be the reaction go to 100% completion and will no longer be reversible.

The only thing that will change the equilibrium situation is if something occurs to "mess up" the newly established equilibrium. Since the rates of the forward (and reverse) processes depend on the amounts (concentration) of the solid (or ions) present, equilibrium is disturbed by either:

- Adding more reactant or product from another source to the solution; or
- Removing some of the existing reactant or product from the solution.

For example, if some NaCl was added to the to the $PbCl_2$ equilibrium mixture (which, as mentioned, contains some solid $PbCl_2$ along with some dissolved Pb^{2+} and Cl^- ions), the NaCl would dissociate in water according to Eq. 3:

$$NaCl(s) \xrightarrow{\ H_2O\ } Na^+(aq) + Cl^-(aq) \qquad [3]$$

This means that adding NaCl to the equilibrium mixture is equivalent to adding more Cl^- ions. This increases the likelihood of Pb^{2+} and Cl^- ions recombining and, thus, speeds up the reverse reaction. As a result, additional solid $PbCl_2$ will be produced until there is enough $PbCl_2$ to increase the rate of the forward reaction enough, so the two are once again equal. Since adding more of the product, Cl^-, to the reversible reaction in Eq. 18 resulted in more reactant being produced, we say that **adding a product (Cl^-) shifted the equilibrium to the left** (or **towards reactants**):

$$PbCl_2(s) \rightleftharpoons Pb^{2+}(aq) + 2\,Cl^-(aq)$$

more Cl^- added

increases   decreases

In the same way, removing some Cl^- would result in the opposite type of change—by removing some of the ions, you would reduce the chances of them recombining, lower the speed of the reverse reaction, and cause more solid $PbCl_2$ to break up into ions until enough ions are produced to make the forward and reverse reaction rates equal once again:

$$\overrightarrow{} \quad \boxed{Cl^- \text{ removed}}$$
$$PbCl_2(s) \;\rightleftharpoons\; Pb^{2+}(aq) + 2\,Cl^-(aq)$$

We would say that **removing a product (Cl⁻) shifted the equilibrium to the right** (or **towards products**). The removal of Cl⁻ can be accomplished by adding some other substance that will react with it—for example, a silver nitrate solution ($AgNO_3(aq)$), since silver ions form an insoluble precipitate AgCl with chloride ions:

$$Ag^+(aq) + Cl^-(aq) \;\rightarrow\; AgCl(s) \qquad\qquad [4]$$

One other way to disturb the equilibrium of a reversible reaction is by changing the temperature. This has to do with the fact that all chemical reactions either require (absorb) heat or release heat. in the $PbCl_2$ equilibrium, heat is absorbed to break up $PbCl_2$ into Pb^{2+} and Cl⁻, so we can rewrite Eq. 1 this way:

$$heat + PbCl_2(s) \;\rightleftharpoons\; Pb^{2+}(aq) + 2\,Cl^-(aq) \qquad\qquad [5]$$

Generally, reactions that require (absorb) heat are called **endothermic**:

$$heat + A \;\rightleftharpoons\; B \qquad\qquad [6]$$

and reactions that release heat are called **exothermic**:

$$A \;\rightleftharpoons\; B + heat \qquad\qquad [7]$$

It should be clear from our discussion and from Eq. 5 that raising the temperature of the $PbCl_2$ equilibrium mixture would result in a shift of the equilibrium to the right (since more heat is being added):

$$\boxed{heat\ added} \qquad \overrightarrow{}$$
$$heat + PbCl_2(s) \;\rightleftharpoons\; Pb^{2+}(aq) + 2\,Cl^-(aq)$$

and lowering the temperature is equivalent to removing heat, shifting the equilibrium to the left:

$$\boxed{heat\ removed} \qquad \overleftarrow{}$$
$$heat + PbCl_2(s) \;\rightleftharpoons\; Pb^{2+}(aq) + 2\,Cl^-(aq)$$

Notice that each time the equilibrium is disturbed, the reaction returns to a condition of equilibrium, although the ratio of reactants : products is different in the new equilibrium—and both reactants and products are always present. One way to summarize the "response" of the reacting mixture was formulated by a French chemist, Henri LeChâtelier (pronounced "le-**SHAH**-tell-YAY"): *If a system* (reaction or physical change) *at equilibrium is disturbed, the position of the equilibrium will shift in the direction opposing the disturbance.* This statement is called **LeChâtelier's principle** in his honor.

In other words, if a reactant is added, increasing its amount, the equilibrium will shift to the right, which has the effect of using up the "extra" reactant. If a product is removed, it would also result in a shift to the right—the system "responds" to the decreased amount of product by reducing the amount of reactants to increase the amount of products to make up for the removal.

In general, then, according to LeChâtelier's principle:

- If reactants are added (or products are removed), the equilibrium system responds by producing more products and reducing the amount of reactants in the mixture;

- If products are added (or reactants are removed), the equilibrium system responds by producing more reactants and reducing the amount of products in the mixture.

In this experiment, you will examine two equilibrium reactions and explain their behavior when different disturbances are introduced.

Objectives

1. To observe reversible chemical reactions and equilibrium disturbances
2. To identify and discuss factors that cause a shift in the equilibrium
3. To identify reactions as endothermic or exothermic

Safety

- Wear approved **safety glasses or goggles** during the entire experiment!

- Avoid **burns** by contact with hot objects (hot plate, boiling water)! Never hand hot objects to another person!

- Many of the chemicals used in the demonstration and experiment are potentially harmful! Prevent contact with your eyes, skin, or clothing. Avoid swallowing any of the reagents. Specifically:

  o Concentrated hydrochloric acid (12 M HCl) is toxic and corrosive! It can cause severe burns, especially if you get it in your eyes!

  o Cobalt(II) chloride ($CoCl_2$) dissolved in 95% ethanol is flammable, toxic, and irritating! Wash it off your skin immediately if spilled.

  o Saturated ammonium chloride solution is an irritant! Wash it off your skin immediately if spilled.

- Wash your hands thoroughly with soap before you leave the lab room!

Materials and Equipment

You will need the following for your experiment:

1. Cobalt(II) chloride solution (in 95% ethanol)
2. Ammonium chloride solution, saturated
3. Hydrochloric acid, concentrated
4. Ice (for an ice bath)
5. Test tubes (2)
6. Hot plate
7. Beakers, 250 mL and 400 mL
8. Glass stirring rod
9. Plastic pipette

Procedure

I. Cobalt ions

1. Set up a hot water bath and an ice bath:

   - Half-fill the 400-mL beaker with warm water from the faucet and set it on the hot plate. Set the HEAT knob on the hot plate to 300°C.

   - Use ice from the communal ice bucket to half-fill the 250-mL beaker with ice. Add a small amount of *cold* water from the faucet.

2. Obtain a pipetteful of the cobalt(II) chloride solution in a test tube. Record the initial color of the solution on the Data/Observation Sheet.

3. Add water to the solution dropwise by using your plastic pipette until you observe a significant change of color. *You may need to stir the solution with your glass stirring rod as you are adding this and other solutions.*

   Record the new color on the Data/Observation Sheet.

4. Use the pipette that attached to the bottle of concentrated hydrochloric acid to add HCl to the solution dropwise until you observe a significant change of color. Record the new color on the Data/Observation Sheet.

   > **_IMPORTANT_:** *Be careful not to add too much acid! This may keep you from successfully completing the next steps. If you have added 10 drops and there is still no color change, stop and get further instructions from the professor.*

5. Place the test tube in the ice bath and allow it to sit there for a few minutes. While you are waiting, begin doing Part II of the Procedure (the ammonium chloride solution). Stir occasionally by swirling the test tube in the beaker. Once the color of the solution changes, record the new color on the Data/Observation Sheet.

6. Take the test tube out of the ice bath and place it in the hot water bath. Allow the test tube to sit in the hot water bath for a few minutes. Stir occasionally by swirling the test tube in the beaker.

7. Once the color change has occurred, record the new color on the Data/Observation Sheet.

II. Ammonium chloride

8. Obtain a pipetteful of the ammonium chloride solution in a test tube. Record the initial appearance of the solution on the Data/Observation Sheet.

9. Use the pipette that came with the bottle of concentrated hydrochloric acid to add HCl to the solution dropwise until you observe a significant change. Record your observations on the Data/Observation Sheet.

10. Place the test tube in the hot water bath. Once a change in the solution occurs, record the new color on the Data/Observation Sheet.

11. Take the test tube out of the hot water bath and place it in the ice bath. Allow the test tube to sit in the ice bath for a few minutes. Stir occasionally by swirling the test tube in the beaker.

12. Once a change in the solution occurs, record the new color on the Data/Observation Sheet.

13. Before you dispose of your ammonium chloride solution, simply place your solution back in the hot water bath and allow it to sit there until all of the solid is dissolved!

Clean-Up

- Discard the cobalt(II) chloride solution in the "USED $CoCl_2$ SOLUTION" at the front desk.

- Discard the ammonium chloride solution in the "USED NH_4Cl SOLUTION" at the front desk. Don't forget to heat the solution until all of the precipitate has dissolved!

- Wash any test tubes you used and place them back into your equipment drawer.

- Clean up chemical spills with a wet paper towel; dry the benchtop with a dry paper towel.

- Pour out the ice bath and hot water bath down the drain. Wash your hands before leaving!

Results/Conclusions

Use your observations from the Data/Observation Sheet to explain how the equilibrium shifts in response to the disturbances that you created (adding other reagents, heating, cooling, etc.). Record your conclusions on the Result/Conclusion Sheet. Be sure to explain briefly the **evidence** you observed that confirms your statements, namely:

- How the color (or appearance) change observed corresponds to the chemical equation
- What causes the shift in equilibrium—and how it fits with LeChâtelier's principle (whether it is addition or removal of a reactant or product, including heat)

More specific guidelines are listed below for each item on the Result/Conclusion Sheet.

1. Compare the color change after the addition of H_2O to the colors of the $[CoCl_4]^{2-}$ and the $[Co(H_2O)_6]^{2+}$ ions in the equation given on the Result/Conclusion Sheet. What does that tell you about which way the equilibrium shifted (left = to the reactants, or right = to the products)? Also tie this to what you are adding to the mixture (H_2O) and whether it appears somewhere in the reaction.

2. Something that will help you decide what the effect of the addition of HCl (besides the obvious color change) is the fact that HCl, when it is dissolved in water, will actually break up into H^+ and Cl^- ions (Eq. 8):

$$HCl(aq) \rightarrow H^+(aq) + Cl^-(aq) \qquad\qquad [8]$$

   Look carefully at the ions produced! Is any of them a part of the original equilibrium between the cobalt(II) ions? This should give you a clue!

3. You do not know up front whether heat is a reactant or product—in fact, the whole point here is to find out this information and identify the reaction as either exothermic or endothermic! To do this, try putting heat as a reactant first, then as a product, into the equilibrium reaction equation and see what effect the addition of heat will have. Then compare your prediction of which way the equilibrium would shift to which way it *actually* shifted. Either the endothermic or the exothermic option will match your observations.

4. This simply asks you to write "**+ heat**" on either the reactant or the product side of the equation listed on the Result/Conclusion Sheet in Part I. Do this according to your answer in Step 3 above.

5. Consider the same effect that you looked at for the cobalt(II) equilibrium in Step 2 above.

6. Use the same considerations you did in Step 3 above.

7. Add "**+ heat**" on either the reactant or the product side of the equation listed on the Result/Conclusion Sheet in Part II—exactly like you did in Part I under Step 4.

_______________________________   __________   __________
name   section   date

Pre-Laboratory Assignment

1. Define the following terms, as they apply to this experiment:

    a.   Reversible reaction –

    b.   Chemical equilibrium –

    c.   LeChâtelier's principle –

    d.   Equilibrium disturbance –

2. Which way will the equilibrium shift (left or right) in the reaction below:

$$Fe^{3+}(aq, \textbf{yellow}) + SCN^{-}(aq) \rightleftharpoons FeSCN^{2+}(aq, \textbf{red}) + heat$$

    a.   When more SCN^{-} ion (in the form of KSCN) is added to the reaction mixture? *Explain why.* *HINT:* Addition of reactants shifts equilibrium toward product; addition of product shifts equilibrium toward reactant.

b. When SCN^- ion is removed from the reaction mixture (by adding Ag^+ ions, which form an insoluble AgSCN(s) precipitate and thus reduce the amount of SCN^-)? *Explain why.*

c. When the temperature of the mixture is increased? *Explain why.*

d. Would adding more Fe^{3+} result in the color change to yellow or red? *Explain why.*

name section date

Data/Observation Sheet

I. Cobalt ions

Initial color of equilibrium mixture ______________

Color after addition of water, H_2O ______________

Color after addition of hydrochloric acid, HCl ______________

Color after cooling ______________

Color after heating ______________

II. Ammonium chloride

Initial appearance of solution ______________

Appearance after addition of hydrochloric acid, HCl ______________

Appearance after heating ______________

Appearance after cooling ______________

____

name   section   date

Result/Conclusion Sheet

I. Cobalt ions

$$[CoCl_4]^{2-}(aq, blue) + 6\ H_2O(\ell) \rightleftharpoons [Co(H_2O)_6]^{2+}(aq, pink) + 4\ Cl^-(aq)$$

a. Which way did the equilibrium shift (left or right) after the addition of H_2O and why? (*Explain!*) *HINT:* Equilibrium shifts to right (toward product) with the addition of reactants; equilibrium shifts to the left (toward reactant) with the addition of product.

b. Which way did the equilibrium shift after the addition of HCl and why? (*Explain!*)

c. Which way did the equilibrium shift after the mixture was heated? Is the reaction exothermic (heat is a product) or endothermic (heat is a reactant)?

d. Write the **heat** term on the correct side of the chemical equation above, according to your answer to Question 3.

II. Ammonium chloride

$$NH_4Cl(s, white) \rightleftharpoons NH_4^+(aq) + Cl^-(aq)$$

4. Which way did the equilibrium shift after the addition of HCl and why? (*Explain!*)

5. Which way did the equilibrium shift after the mixture was heated? Is the reaction exothermic or endothermic?

6. Write the **heat** term on the correct side of the chemical equation above, according to your answer to Question 6.

_______________________________________   __________   __________
name   section   date

Post-Laboratory Assignment

1. When a reaction system reaches chemical equilibrium…

    a. What happens to the rates of the forward and reverse reactions? *Explain.*

    b. What happens to the amounts of reactants and products? *Explain.*

2. For the **exothermic** reversible reaction below at equilibrium:

$$2\ H_2(g) + O_2(g) \rightleftharpoons 2\ H_2O(g)$$

    a. How would the amount of O_2 change when H_2 is added to the reaction mixture?

    b. How would the amount of H_2 change when H_2O is added to the reaction mixture?

    c. How would the amount of H_2O change when temperature is decreased?

Mike Goldin (Liberty University)

They also gave me gall for my food, and for my thirst they gave me <u>vinegar</u> to drink.
– Psalm 69:21 (NKJV)

Background

A **neutralization reaction** is a type of double-displacement reaction, where one reactant is an acid, the other is a base, and the two products are water and a **salt**. An example of this is the reaction between a solution of vinegar (containing acetic acid, $HC_2H_3O_2$) and a solution of sodium hydroxide (NaOH, which is a base), shown in Eq. 1:

$$NaOH(aq) + HC_2H_3O_2(aq) \rightarrow NaC_2H_3O_2(aq) + HOH(\ell) \tag{1}$$

We can use what we know about mass relationships in chemical reactions and extend this to aqueous solutions. If we do not know the concentration of acetic acid in the vinegar, we can add enough of the sodium hydroxide solution (whose concentration we do know) to neutralize it. Since NaOH and acetic acid have a mole ratio of 1:1 from Eq. 11, if we add the number of moles of NaOH exactly equal to moles of acetic acid in the vinegar, our solution should end up approximately neutral. We can then use the known concentration of NaOH and its volume, as well as the volume of $HC_2H_3O_2$, to calculate the molarity of acetic acid.

This process of measuring the volume added of the reagent (whose molarity is known) to a known volume of another solution (whose molarity is unknown) in order to calculate the unknown molarity is called **titration**. We usually add the known reagent to the one being measured by using a **burette** (also spelled **buret** and pronounced "**byoo-RETT**" and not "**buh-RAY!**"), which is a glass tube with volume markings on its side and a **stopcock** at the bottom to start and stop the flow of liquid from it. A typical titration setup, like the one you will use in this experiment, is shown in Figure 1.

A burette differs from other volumetric glassware that you have seen so far, such as a graduated cylinder, in that the reading at the top of the burette will always read, "0 mL," as opposed to the highest volume (in this case, 50 mL). In fact, the 50-mL graduation is the one at the bottom of the burette, and there is also some unmarked/ungraduated space below that mark. What this means is that the amount of the solution that is inside the burette cannot be determined. In fact, the point of the burette is to tell us how much solution we have dispensed by recording the level (milliliter mark) before and after it is dispensed and then subtracting the initial number from the final to get the actual volume dispensed.

However, another problem is how to know that enough NaOH was added. Both NaOH and acetic acid are clear, colorless solutions! The answer to this question lies in adding an **acid-base indicator** to the reaction mixture. A very useful, commonly used indicator

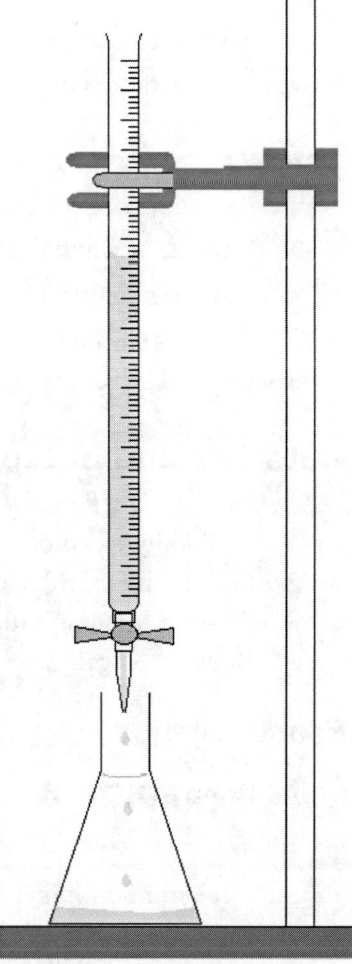

<u>Figure 1</u>. Titration setup

for acid-base titrations is **phenolphthalein**. If one or two drops of phenolphthalein is added to the acid solution and then *titrate* the acid with the NaOH, the solution will stay colorless as long as its pH is below 8.2 (see Table 2 in Experiment 11). In other words, as long as the solution is acidic, or even when equal amounts of acid and base are present (and the solution is at about neutral pH), it will still remain colorless.

However, the *very next drop* of NaOH makes the solution basic! At that point, you see a very faint pink color as the pH range where the phenolphthalein color changes from colorless to pink is reached. This is called the **end point** of the titration and is the signal to stop adding NaOH, as enough to make the solution neutral, plus one more drop was added. The error from one drop is so small that we simply record the volume of NaOH added without worrying about the extra drop that was added *after* all of the acid had reacted.

In this experiment, the effectiveness of one of the common brands of antacids (such as Alka-Seltzer) will be measured by crushing one tablet, adding enough hydrochloric acid (which is chemically the same as gastric juice) to it for all of the antacid to react with it, and then titrating the remaining acid to determine the difference between how much acid was added and how much actually remains in solution after the reaction with the antacid. Exact steps for the titration process and data/calculations needed to figure out the concentration of the remaining acid, as well as the effectiveness of the antacid, are discussed in "Procedure" and "Calculations."

Objectives

1. To titrate a solution of an acid with a base using a burette and an acid-base indicator
2. To calculate the molarity of an acid from titration data

Safety

- Wear approved **safety glasses or goggles** during the entire experiment! It is especially important when we are working with toxic, corrosive, and irritant substances.
- Aqueous solutions of acids and bases are toxic, corrosive, and can irritate or cause burns.
- Remember the location of the emergency eyewash and safety shower in case of accidents!
- Wash your hands thoroughly before leaving the laboratory.

Materials and Equipment

You will need the following for your experiment:

1. 0.20 M hydrochloric acid for antacid
2. Sodium hydroxide solution for titration
3. Phenolphthalein indicator
4. Burette, 50 mL
5. Beakers (2)
6. Erlenmeyer flask, 125 mL or 250 mL
7. Graduated cylinder, 100 mL
8. Glass funnel (optional)

Procedure

I.   Addition of HCl to the crushed antacid

1. Weigh a clean, dry 250-mL Erlenmeyer flask. Record its mass on Line 1 of the Data Sheet.
2. Choose the antacid brand you wish to test, and record the brand name on Line 2 of the Data Sheet. Take one tablet of your chosen brand and crush it by placing it in a folded piece of paper. Then use a spoon or porcelain pestle to apply pressure and crush the tablet.

3. Transfer the crushed tablet into an Erlenmeyer flask, taking care not to spill any of the powder. Record the combined mass of the crushed tablet and flask on Line 3 of the Data Sheet.

4. Measure precisely 30.0 mL of 0.20 M hydrochloric acid in a 100-mL graduated cylinder and add the acid to the powdered tablet in the flask. Stir periodically with a glass rod or by swirling the liquid in the flask. *You should observe some bubbling when acid is added.*

5. ***VERY IMPORTANT!!!*** Once the bubbling has stopped, add 2-3 drops of phenolphthalein indicator to the reaction mixture in the Erlenmeyer flask. You should have added enough acid to completely react with all of the antacid, so the pH of the solution should be below 8.2, meaning the phenolphthalein should stay colorless.

   If it turns the solution pink, there was not enough acid added! In this case, add an extra 10.0 mL of 0.2 M hydrochloric acid to the mixture, which should turn it colorless. If not, continue adding 10.0-mL portions of hydrochloric acid until the pink color disappears.

6. After any additional hydrochloric acid has been added, record the total volume of the acid added to the antacid on Line 5 of the Data Sheet – this will be either the 30.0 mL as directed in Step 4, or 40.0 mL, 50.0 mL, etc., if you had to add more acid in Step 5.

   You are now ready to begin titrating the hydrochloric acid to find out the amount of excess acid remaining.

II. Titration of remaining acid with NaOH

7. To begin titrating, you first have to fill your burette with the sodium hydroxide solution provided for the titration.

   First, turn the burette over, so that its tip is pointing downward: loosen the clamp holding the burette, **carefully** slide it out, and reinsert and clamp the burette. Set aside the small beaker that sits underneath the burette to catch the dripping water.

8. Place a "discard" beaker from your drawer (which does not have to be clean/dry) underneath the burette to catch any spillage, as well as some NaOH you drain out later.

9. Pour about 50-60 mL of sodium hydroxide labeled *"For titration"* into another **clean, dry** beaker. Record the molarity of the NaOH written on the bottle on Line 6 of your Data Sheet. Use this solution to fill your burette. (You may want to lower the burette and to insert a glass funnel in the top of the burette to help you fill it without spilling.)

   > **IMPORTANT**: *Clean up any spills immediately by using paper towels to soak in the spilled NaOH, then washing off the benchtop with plenty of water to prevent **permanent damage** to the benchtop! Ask the instructor for help if necessary.*

10. Fill the tip of the burette with the NaOH solution by opening the stopcock for a few seconds and allow a small amount of the solution to drain from the burette into the discard beaker. Close the stopcock!

11. Record the **exact** initial volume reading on the "NaOH burette" on Line 7 of the Data Sheet, **using rules of significant figures**. (*HINT: The initial reading doesn't have to be 0.00 mL!*)

12. Move the discard beaker out of the way and place the Erlenmeyer flask with the antacid/HCl mixture directly under the burette tip. If necessary, move the burette down in the clamp until the tip is inside the mouth of the flask.

   HINT: Place a white sheet of paper underneath the flask to help see the color change.

13. Open the stopcock and add 1-2 mL of NaOH to the solution. Close the stopcock and swirl the flask. Repeat the addition of 1-2 mL at a time until you begin to see a small plume of pink in the solution inside the flask.

14. Once the colored plume begins to appear, stop adding NaOH and swirl the flask. The color will them disappear. Add a few more drops of NaOH at a time by only partially opening the stopcock. Swirl the flask to mix the solution.

    HINT: Another way to do this is to quickly spin the stopcock around 180°, which delivers a small burst into the flask.

15. When the whole solution turns completely light pink after you swirl it, you have reached the endpoint. Stop adding any more NaOH at this time. Record the **exact** final NaOH volume reading on the burette on Line 8 of the Data Sheet, **using significant figure rules**.

Clean-Up

- Clean up any spills by using paper towels to soak in the spilled acid or base, then washing off the benchtop with plenty of water. Do this as soon as you find a spill!

- Drain the remaining NaOH solution from the burette into the original beaker with NaOH. If the solution in this beaker was not contaminated, return the NaOH to the original container. If it was mixed with other solutions or contaminated in any other way, discard it in the "MIXED ACIDS AND BASES DISCARD" container.

- Discard the titrated antacid/HCl solution from the Erlenmeyer flask into the "MIXED ACIDS AND BASES DISCARD" container.

- Clean the burette by filling a clean beaker with tap water and using it to fill the burette. Drain the water from the burette into the discard beaker. Fill the burette with water and drain it once more. Loosen the clamp, **carefully** pull out the burette, and clamp it upside down.

> _IMPORTANT_: **Place the small beaker back underneath the burette to catch the dripping water!**

- Return the glass funnel to the front desk.

Calculations

Use the data from your Data Sheet to do these calculations. Record results under Calculations.

1. Calculate the mass of the crushed antacid tablet by subtracting the mass of the empty Erlenmeyer flask from the combined mass of the flask and the crushed antacid, as recorded on the Data Sheet. Record this on Line 9 under Calculations.

2. Calculate the volume of NaOH used (added) to titrate the vinegar by subtracting the initial reading on *your* burette from the final reading (Eq. 2). Record this on Line 10 under Calculations.

$$V_{\text{NaOH}} = V_{\text{final}} - V_{\text{initial}} \qquad [2]$$

3. Calculate the volume of the portion of hydrochloric acid that did not react with the antacid—this is the excess acid remaining. ***This is the most complex part of the calculations!*** First off, you need to know the chemical equation for the reaction of NaOH and HCl, which is shown below (Eq. 3). Record this on Line 11 under Calculations.

$$\text{NaOH(aq)} + \text{HCl(aq)} \rightarrow \text{NaCl(aq)} + \text{HOH}(\ell) \qquad [3]$$

From Eq. 3, you can get the mole ratio of HCl to NaOH. The volume of NaOH added was just calculated according to Eq. 2. Finally, to convert between moles and volumes of NaOH and HCl, you would need to use their molarities (both of which are known), remembering that the notation "M" represents "mol/L." The general way to do this is shown below (Eq. 4), where filled-in squares "■" represent numbers you have from other data or calculations in this experiment. (This is very similar to mass relationship calculations!)

$$V_{\text{excess HCl}} = \blacksquare \text{ L NaOH} \times \frac{\blacksquare \text{ mol NaOH}}{1 \text{ L NaOH}} \times \frac{\blacksquare \text{ mol HCl}}{\blacksquare \text{ mol NaOH}} \times \frac{1 \text{ L HCl}}{\blacksquare \text{ mol HCl}} \qquad [4]$$

*HINT: Note the difference in volume units—in lab, we measure in **milliliters**, while molarities are in moles per **liter**!*

4. Now that we know the volume of excess (unreacted) HCl, we can calculate the volume of reacted hydrochloric acid by subtracting it from the volume of acid added at the beginning, which was recorded on Line 5 of your Data Sheet (Eq. 5). Record this on Line 12 under Calculations.

$$V_{\text{reacted HCl}} = V_{\text{added HCl}} - V_{\text{excess HCl}} \qquad [5]$$

5. To estimate the effectiveness of the antacid, we will make two assumptions. The first assumption is that the acid we used (0.20 M hydrochloric acid) has twice the concentration of HCl found in gastric acid of an average person. Since we used the more concentrated acid, it would have taken only **half the amount** of 0.2 M HCl to react with the same amount of antacid as it would take of the 0.1 M HCl.

Therefore, calculate the volume of 0.10 M HCl, or "gastric acid," equivalent to the actual volume of 0.20 M HCl by multiplying $V_{\text{reacted HCl}}$ by two (Eq. 6) and record this on Line 13 under Calculations.

$$V_{\text{"gastric acid"}} = 2 \times (V_{\text{reacted HCl}}) \qquad [6]$$

6. The second assumption we must make is that the density of "gastric acid" is roughly equal to the density of pure water, or 1.00 g/mL. This would allow us to convert milliliters of "gastric acid" to grams.

Using this assumption for density, calculate the mass (grams) of "gastric acid" based on the volume (mL) calculated in Step 5. Record this on Line 14 under Calculations.

7. Finally, to find the effectiveness of the antacid, we should calculate how much acid would be neutralized by exactly one gram of the antacid. This way, no matter how big or small the tablet—or how many tablets the patient takes—you would have a way to compare different antacids "apples to apples!"

Calculate the ratio of the mass of acid neutralized to the mass of antacid by dividing the mass of the "gastric acid" from the last calculation step over the mass of the crushed antacid tablet that was calculated in Step 1 (as shown in Eq. 7 below):

$$\text{Antacid effectiveness} = \frac{\text{mass of "gastric acid" from Step 6}}{\text{mass of crushed antacid tablet from Step 1}} \qquad [7]$$

This calculated number represents the effectiveness of the antacid – how many grams of acid it can neutralize per gram of the antacid taken by the patient. Record this on Line 15 under Calculations.

---

name	section	date

Pre-Laboratory Assignment

1. Define these terms (some of which you may need to look up in previous labs):

    a. Neutralization reaction –

    b. Acid-base indicator –

    c. Molarity –

    d. Titration –

    e. End point –

2. What are differences between a graduated cylinder and a burette? *Name at least two!*

3. a) How would you record the following volume measurement on a burette shown in the picture using rules of significant figures?

$V =$ ___________________ mL

b) If the burette was used to dispense a certain amount of solution, and the new volume reading is shown in the picture below, what volume of the solution was dispensed? *Use rules of significant figures and show your work!*

$V_{dispensed} =$ ___________________ mL

Show work:

______________________________   ______   ______
name   section   date

Data Sheet

1. Mass of empty Erlenmeyer flask, g   ______________________

2. Antacid brand used   ______________________

3. Combined mass of crushed tablet and flask, g   ______________________

4. Molarity of HCl, M   ______________________

5. Total volume of HCl added to antacid, mL   ______________________

6. Molarity of NaOH (from the bottle), M   ______________________

7. Initial burette reading of NaOH, mL   ______________________

8. Final burette reading of NaOH (at the endpoint), mL   ______________________

Calculations

The steps to calculate these values can be found on page 136 and 137.

9. Mass of the crushed antacid tablet, g   ______________________

10. Volume of NaOH added, mL   ______________________
    Show work.

11. Volume of excess 0.20 M HCl, mL   ______________________
    Show work:

12. Volume of 0.20 M HCl reacted with antacid, mL   ______________________
    Show work

13. Equivalent volume of "gastric acid" (**0.10 M** HCl), mL ______________________
    Show work

14. Mass of 0.10 M HCl "gastric acid" reacted with antacid, g   ______________________

15. Mass of "gastric acid" neutralized by 1 gram antacid, g   ______________________
    Show work:

name   section   date

Post-Laboratory Assignment

1. Consider these experimental mistakes in the titration of the antacid/HCl mixture with sodium hydroxide:

   a. What would happen if you forgot to add the phenolphthalein to the antacid/HCl mixture? *Explain!*

   b. The burette had 10 mL of water in it before it was filled with NaOH. Would this increase, decrease, or not affect the molarity of NaOH in the burette? *Explain!*

   c. Based on your answer to part (b), would the 10 mL of water increase, decrease, or not affect the **calculated** effectiveness of the antacid? *Explain!*

2. A student was titrating pure hydrochloric acid of unknown molarity. He forgot to take the initial volume reading for the NaOH addition in his buret and decided to assume it was 0.00 mL. His final reading on the NaOH buret was 23.45 mL, and the volume of hydrochloric acid used for titration was 25.00 mL. The molarity of the NaOH is 0.120 M.

    a. Calculate the molarity of HCl based on his data. Show your work: (*HINT:* This is a very similar procedure for calculating excess HCI acid. Remember for the last step that molarity equals mol/L).

    b. His lab partner then remembers that the *actual* initial volume of NaOH in the buret was 5.70 mL! Now recalculate the molarity of HCl based on this *real* number!

    c. Considering the calculation in part (b) the "true" molarity, calculate the percent error in the molarity calculation:

$$\% \text{ error} = \frac{\text{true molarity} - \text{calculated molarity}}{\text{true molarity}} \times 100\%$$

    d. Was it okay for the student to assume that the initial volume reading of NaOH was 0.00 mL? *Explain!*

EXPERIMENT 9: **Naming Organic Compounds**

Randy Davy, Mark Hemric, Mike Goldin (Liberty University)

*Out of the ground the LORD God formed every beast of the field and every bird of the air,
and brought them to Adam to see what he would call them.
And whatever Adam called each living creature, that was its name.*
– Genesis 2:19 (NKJV)

Background

For detailed background information, please read Chapter 6 of the Guinn/Brewer textbook (pp. 179-230). The following is a brief summary of the most important points from the text and lectures.

Nail polish remover, vinegar, and propane in your grill lighter are three common "everyday" substances. The first one is a liquid, the second, an acid (dissolved in water), and the third one a gas. Despite their differences, all three belong to a large number of chemical compounds called **organic compounds. Organic chemistry** is a branch of chemistry that studies compounds made up primarily of carbons and hydrogens. The name "organic" comes from the fact that living things (animals, plants, man), or *organisms*, are made of similar compounds. The simplest organic compound is **methane**, better known as natural gas, whose Lewis structure and (tetrahedral) molecular shape are shown below:

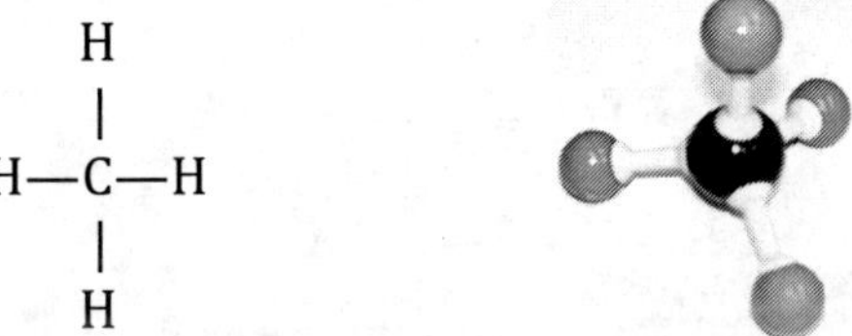

There are actually millions of organic compounds currently known to man, many of which are man-made. We will spend a considerable amount of time this semester discussing different aspects of organic compounds. This lab activity is an introduction to their naming and classification.

The main part of organic compounds is a **carbon chain**—a number of carbon atoms bound to each other with nonpolar covalent bonds. The name of most organic compound contains a "root" part that tells how many carbons are in it. Table 1 shows the most common root names.

<u>Table 1</u>. Root names for organic compounds

# Carbons	Root Name (Pronunciation)	
1	meth-	
2	eth-	
3	prop-	
4	but-	(pronounced BEW-T)
5	pent-	
6	hex-	
7	hept-	
8	oct-	
9	non-	
10	dec-	(pronounced DECK; sometimes DESS)

The simplest type of organic compounds is **hydrocarbons**, which contain only carbon and hydrogen atoms. Those hydrocarbons with only single C—C bonds are called **alkanes** and will have the ending **–ane** to their names. We can use different ways to represent organic molecules:

- **Molecular formulas**, with information only about the # atoms of each element
- **Expanded structural formulas**, which are Lewis structures without lone pairs
- **Condensed structural formulas**, where C—H bonds are not shown
- **Skeletal line structures** (or **geometric formulas**), where H's are omitted and each C is shown as a vertex or end of a line

Table 2. Structures (depictions) of organic compounds

Name	Molecular Formula	Expanded Struc. Formula	Condensed Structural Formula / Skeletal Line Structure
methane	CH_4	H—C—H (with H above and below)	CH_4 (*no line structure*)
ethane	C_2H_6	H—C—C—H (expanded)	$CH_3—CH_3$
propane	C_3H_8	H—C—C—C—H (expanded)	$CH_3—CH_2—CH_3$
butane	C_4H_{10}	H—C—C—C—C—H (expanded)	$CH_3—CH_2—CH_2—CH_3$
pentane	C_5H_{12}	H—C—C—C—C—C—H (expanded)	$CH_3—CH_2—CH_2—CH_2—CH_3$
hexane	C_6H_{14}	H—C—C—C—C—C—C—H (expanded)	$CH_3—CH_2—CH_2—CH_2—CH_2—CH_3$

Hydrocarbons with double C=C bonds are called **alkenes** and will have the ending **–ene** to their names. Those with triple C≡C bonds are called **alkynes** and will have the ending **–yne** to their names.

$$CH_3—CH_3 \qquad \text{eth}\textbf{\textit{ane}}$$
$$H_2C = CH_2 \qquad \text{eth}\textbf{\textit{ene}}$$
$$HC \equiv CH \qquad \text{eth}\textbf{\textit{yne}}$$

Some hydrocarbons are **open chains**, with a beginning and end; in others, the carbon chain is closed into a closed **ring** or **cyclic structure**. (We usually draw cyclic compounds as a line structure—it's much easier!)

$$CH_3—CH_2—CH_2—CH_2—CH_2—CH_3$$ hexane

$$\begin{array}{ccc} H_2C & — & CH_2 \\ / & & \backslash \\ CH_2 & & CH_2 \\ \backslash & & / \\ H_2C & — & CH_2 \end{array}$$ *cyclo*hexane

Isomerism

For hydrocarbons with four (or more) carbon atoms, there is more than one way to arrange them! For example, butane can be arranged in two ways:

$$CH_3—CH_2—CH_2—CH_3 \qquad\qquad CH_3—CH—CH_3$$
$$| $$
$$CH_3$$

These two molecules have the *same number of C and H atoms, but a different arrangement of the atoms.* Such molecules are **isomers** of each other, from the Greek *isos* (equal) and *meros* (part). There are different types of isomers; the two above are called **structural** (or **constitutional**) **isomers**. Below is another example of a type of isomers:

$$\begin{array}{cc} H & H \\ \backslash & / \\ C = C \\ / & \backslash \\ H_3C & CH_3 \end{array} \qquad\qquad \begin{array}{cc} H_3C & H \\ \backslash & / \\ C = C \\ / & \backslash \\ H & CH_3 \end{array}$$

$$\textit{cis} \qquad\qquad\qquad\qquad \textit{trans}$$

In these, the atoms are attached in the same way, except that in the molecule on the left the two ends of the four-carbon alkene chain are both below the C = C double bond, while in the one on the right one part of the chain is above and the other, below the C = C. Isomers like these, where attachments of atoms are the same, but their positioning in space is different, are called **stereoisomers** (from the Greek "solid," or "solid-like" in reference to taking up space).

The two structures above are permanently "locked in" these positions for the carbon chain because it is impossible to turn the carbons on the ends of the *double bond*—this would result in breaking one of the bonds in the double bond! (By contrast, any atom with only a *single bond* to another atom can freely turn about that bond.) This particular type of stereoisomers is called **cis/trans isomers**, with *cis* being the structure with the carbons on the same side and *trans*, on opposite sides of the C = C.

Also, notice that the atoms attached to the ends of C = C are written with every attached atom shown separately, including hydrogens—like in the expanded structural formula; but other parts of the molecule are written as condensed. This way of writing alkene structures (called **bowtie structures** for obvious reasons) is on purpose, to show which parts of the structure are up and which ones are down. If we wrote *either* of the above structures as a strictly condensed formula, we wouldn't know whether it's *cis* or *trans*:

$$CH_3—CH=CH—CH_3$$

Naming Rules

Because we must distinguish between isomers, we must name them differently! But how do we name these two butanes differently?

$$CH_3-CH_2-CH_2-CH_3 \qquad\qquad CH_3-CH-CH_3$$
$$| $$
$$CH_3$$

In order to distinguish between constitutional isomers, we identify them by the longest continuous chain of carbons they have. We call this the **main** (or **parent**) **chain**. (There are a few more details to how the parent chain should be identified. We will discuss them below.)

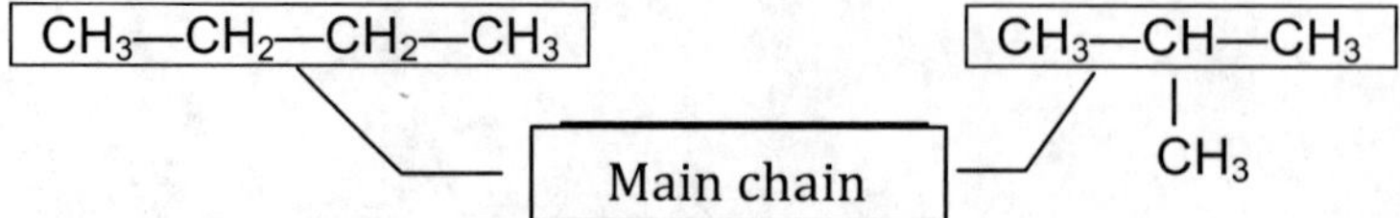

Notice that hydrogens are included as part of the carbon chain! This is because hydrogens are not considered very important when we name organic compounds—they are treated almost like "fillers," filling up the needed bonds so carbon can have four bonds (an octet of electrons). Any part of the molecule—other than hydrogens—that is not part of the parent chain is considered a **substituent**. This term means that the "side chain" of carbons or another atom or group of atoms are there in place of (or *substituting*) a hydrogen atom. For example, the one-carbon side "chain" is a substituent here:

Carbon "side chains" are called "alkyl groups" because their names are the same as alkanes, except the end **–ane** is changed to **–yl**:

Alkane		Alkyl	
CH_4	meth**ane**	CH_3-	meth**yl**
CH_3CH_3	eth**ane**	CH_3CH_2-	eth**yl**
$CH_3CH_2CH_3$	prop**ane**	$CH_3CH_2CH_2-$	prop**yl**

And, of course, since alkyls are attached to the main chain, they must be missing one H atom compared to the alkane with the same name.

Atoms and groups of atoms other than carbons and hydrogen can be attached to a carbon chain as substituents. Such "other" atoms are called **heteroatoms** in contrast to C and H, which will always be found in an organic compound. Some of these drastically change the physical and chemical properties of the organic compound and are called **functional groups**. We will discuss naming compounds with functional groups in detail below, since some special naming rules will apply. The general rule for naming heteroatom substituents is to drop the ending of the element name and use the "stem part" of the name with the ending **-o**.

Substituent	
–Br	brom**o**
–Cl	chlor**o**
=O	**oxo**

When we name compounds, we will name off the main chain and all the substituents, as well as indicate where those substituents are located on the parent chain. The rules below were created by IUPAC (International Union on Pure and Applied Chemistry). Follow these steps:

1. Find the main chain, which is the longest continuous chain of carbons. (Be careful to choose the longest chain, even if it is not written straight across. This chain should contain the most important feature of the molecule, such as a double C = C or triple C ≡ C bond.)

2. Give the parent name (or "last name") based on the number of carbons in the main chain.

   Example:

   Parent or "last" name = **pentane**

3. Give the *name* and *address* of all substituents attached to the parent chain (alphabetically!).

   In our example the substituent is a *methyl* side chain:

   So this is a *methyl*pentane. Now we must give the address of the methyl group; that is we number the main chain and note the number of the carbon to which the methyl group is attached: **3-methylpentane**.

We number the main chain from the end closest to the substituents, or if more than one substituent is on the chain, number from the end that gives the lowest numbers for the addresses. Notice that we use commas between numbers and a hyphen between a number and a word.

EXAMPLE 1

Name the following hydrocarbon:

1. Find the main (parent) chain:

(*continued on the next page*)

EXAMPLE 1 (continued)

2. Assign the parent name: **pentane**

3. Find subsitituents:

$$\text{methyl} \overbrace{CH_3}$$
$$CH_3-CH_2-CH-CH-CH_3$$
$$CH_2-CH_3$$
$$\text{ethyl}$$

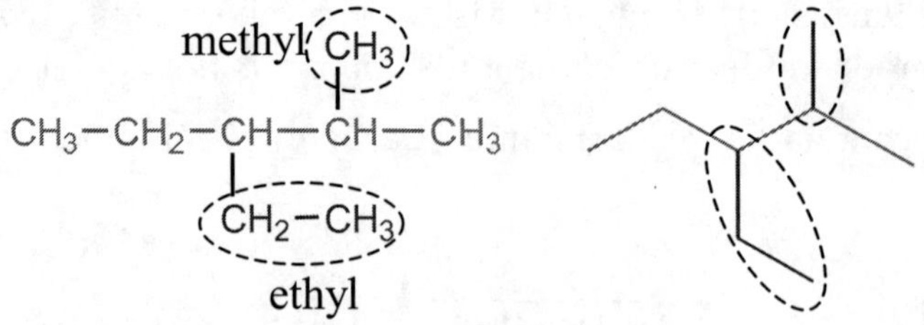

So this will be an ethyl methyl pentane, but we must number the main chain, in this case from right to left:

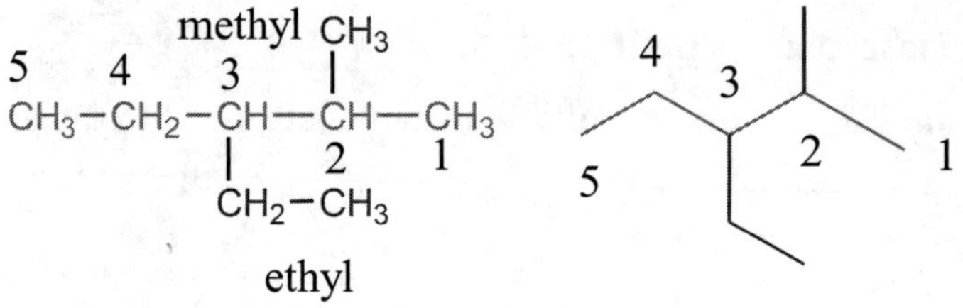

$$\text{methyl } CH_3$$
$$\overset{5}{CH_3}-\overset{4}{CH_2}-\overset{3}{CH}-\overset{2}{CH}-\overset{1}{CH_3}$$
$$CH_2-CH_3$$
$$\text{ethyl}$$

It is **3-ethyl-2-methylpentane**

Here is another example:

EXAMPLE 2

Name the following halogenated hydrocarbon:

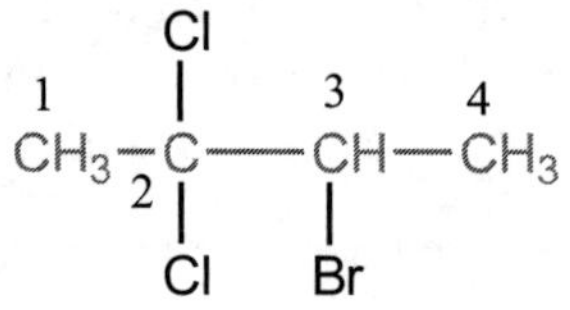

$$Cl$$
$$\overset{1}{CH_3}-\overset{2}{C}-\overset{3}{CH}-\overset{4}{CH_3}$$
$$Cl \quad Br$$

1. Find the main chain:

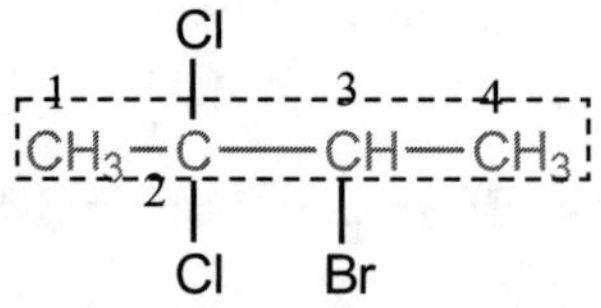

$$Cl$$
$$\overset{1}{CH_3}-\overset{2}{C}-\overset{3}{CH}-\overset{4}{CH_3}$$
$$Cl \quad Br$$

2. Assign the parent name: **butane**

3. Find subsitituents:

$$Cl$$
$$\overset{1}{CH_3}-\overset{2}{C}-\overset{3}{CH}-\overset{4}{CH_3}$$
$$Cl \quad Br$$

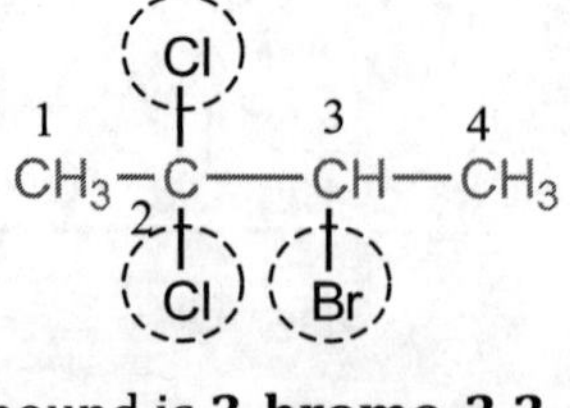

So, the name of this compound is **3-bromo-2,2-dichlorobutane**

Cyclic compounds are named in the same way straight chains are (except, of course, the prefix *cyclo-* for the parent name). Alkenes (C=C) and alkynes (C≡C) are also named in the same way as alkanes, except the following:

- Choose the root name based on the longest continuous carbon chain *that contains both atoms of the double or triple bond*;

- Give the number of the carbon at which the double or triple bond begins (unless obvious);

- Change the ending of the parent name from *-ane* to *-ene* for alkenes (or *-yne* for alkynes);

- For rings that have double or triple bonds, the carbons in the double bond are automatically numbered #1 and #2.

Compounds with Functional Groups

Next, we expand our focus to organic compounds with **functional groups**—substituent atoms or groups of atoms that make the physical and chemical properties of the new compound different from the hydrocarbon parent. When a compound has a functional group, we say that it now belongs to a different "family" of compounds. When we name our new compound, it will also have a different ending to its IUPAC name. For example:

$$CH_3—CH_2—OH \qquad CH_3—CH_2—NH_2 \qquad CH_3—CH=O$$

ethan**ol**   ethan**amine**   ethan**al**

Notice that each of these has a different functional group (circled) and a different ending to their name. The —OH functional group is called **hydroxyl**, the —NH_2 is **amino**, and the = O is a **carbonyl** group. The ending of the compound name reflects the name of the family these compounds belong to: *-ol* for the **alcohol** ethanol, *-amine* for the **amine** ethanamine, and *-al* for the **aldehyde** ethanal. Table 3 below lists the formulas and names for the most common functional groups, the ending of their IUPAC names, and the name of the organic family that has this functional group.

Table 3. Organic "families" and functional groups

Formula	Name	Family	Ending	Remarks
—OH	hydroxyl	Alcohols	-ol	
—SH	thiol	Thiols	-thiol	
—O—	ether	Ethers	ether	*Special naming rules apply*
—N—	amino	Amines	-amine	*May have hydrogens or carbon chains on the N atom*
C=O	carbonyl	Aldehydes	-al	*Carbonyl is on the end*
		Ketones	-one	*Carbonyl is in the middle*
C—O—	carboxyl	Carboxylic acids	-oic acid	*Hydrogen (H) is attached to O—*
		Esters	-oate	*Carbon chain is attached to O—*
C—N—	amido	Amides	-amide	*May have hydrogens or carbon chains on the N atom*

The main differences from hydrocarbon naming include the following:

- The functional group must always be attached to the parent chain—even if this makes it shorter! The carbon with this functional group also gets the lowest number. ***This number usually must be given as part of the name!***

CH₃—CH₂—CH₂—(OH)   CH₃—CH(OH)—CH₃

1-propanol   **2**-propanol

- When naming the parent chain, the hydrocarbon name should have the additional ending (from Table 3).

  Usually the **-e** on the end of the name is dropped before the new ending is added: for example, ethan*e* becomes ethan**ol** or ethan**amine**. However, when the new ending starts with a consonant, the **-e** will not be dropped: ethan<u>e</u>*thiol*.

- For amines and amides, the nitrogen can have carbon substituents on it; when naming the compound, they will have "*N*-" instead of a carbon number in front of the substituent name. For example:

CH₃—CH—CH₂—NH—(CH₃)   *N*,2-<u>dimethyl</u>-1-propanamine
      |
    (CH₃)

- Esters follow a slightly more complicated naming pattern. We will discuss their naming in detail during the laboratory session.

One other family of organic compounds that we will name is **aromatic compounds**. They are also called **benzene derivatives** because they all contain the **benzene** ring shown here:

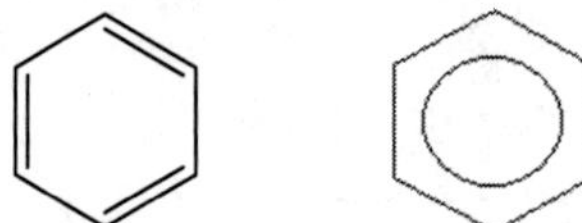

The "donut hole" inside the benzene ring on the right represents the spreading of the alternating single and double bonds, so that each bond can be considered as a "one-and-a-half" bond, not really single or double. This reflects the true nature of the benzene ring.

All aromatic compounds are usually named with benzene as the parent chain:

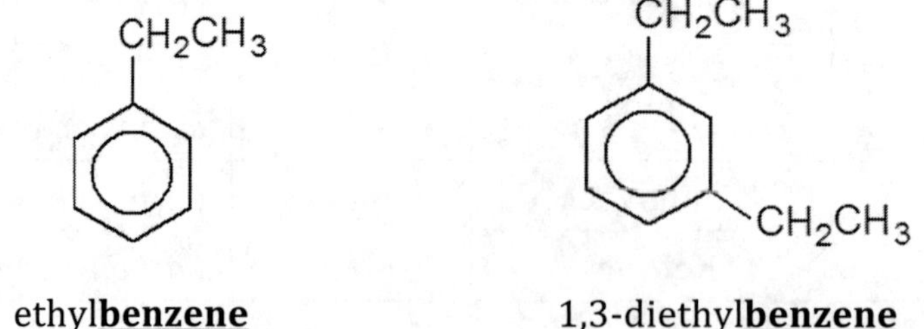

ethyl<u>**benzene**</u>   1,3-diethyl<u>**benzene**</u>

However, sometimes one of the functional groups attached to the benzene ring is named as part of the parent along with the benzene ring. The list of these "common names" for benzene parents is shown in Table 4.

<u>**Table 4.**</u> **Common parent names for aromatic compounds**

Chemical Formula	CH_3	NH_2	OH	$C-H$ (with =O)	$C-NH_2$ (with =O)	$C-OH$ (with =O)
Parent Name	toluene	aniline	phenol	benzaldehyde	benzamide	benzoic acid

During the lab period, we will practice naming many organic compounds as described above. We will also use molecular model kits to help you visualize some molecules.

Objectives

1. Name alkanes, alkenes, alkynes, cycloalkanes, alkyl halides, alcohols, ethers, amines, carboxylic acids, aldehydes, ketones, esters, amides, and aromatic compounds using the IUPAC naming rules
2. Recognize compounds that can have cis/trans isomerism and name them appropriately
3. Properly name an organic compound, given its incorrect name and its structure
4. Draw structures of structural isomers, if the molecular formula of the compound is given
5. Draw structures of alkanes, alkenes, alkynes, cycloalkanes, alkyl halides if their IUPAC name is given

Materials and Equipment

You will need the following for your experiment:

1. A molecular model set

Procedure

Use the IUPAC naming rules for organic compounds to answer the questions on the Worksheet.

name   section   date

Pre-Laboratory Assignment

1. Define the following terms, as they relate to this workshop:

   a. Condensed structural formula –

   b. Bowtie structure –

   c. Structural (constitutional) isomers –

   d. Main (parent) chain –

   e. Functional group –

2. Identify the following compounds as a(n) alkane, alkene, alkyne, cycloalkane, or cycloalkene:

Molecule	Family	Molecule	Family
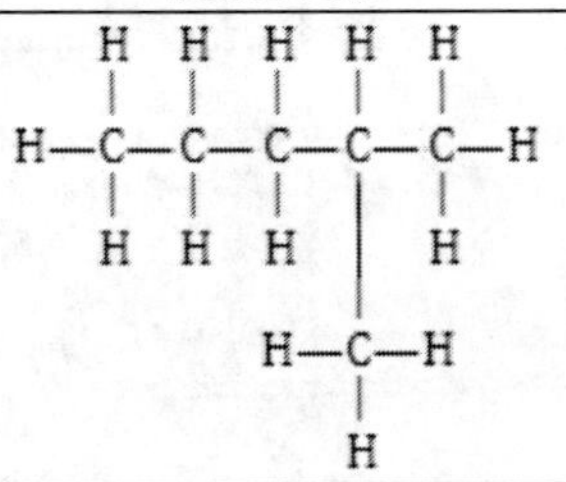 $H-\underset{\underset{H}{\mid}}{\overset{\overset{H}{\mid}}{C}}-\underset{\underset{H}{\mid}}{\overset{\overset{H}{\mid}}{C}}-\underset{\underset{H}{\mid}}{\overset{\overset{H}{\mid}}{C}}-\overset{\overset{H}{\mid}}{\underset{\underset{H-C-H}{\mid}}{C}}-\underset{\underset{H}{\mid}}{\overset{\overset{H}{\mid}}{C}}-H$ (with additional H below)		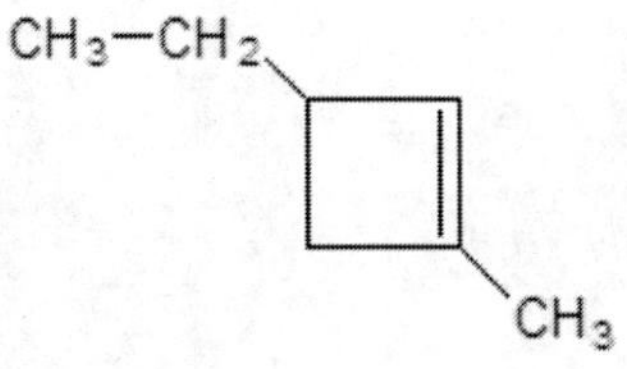 CH_3-CH_2 — (cyclobutane ring) — CH_3	
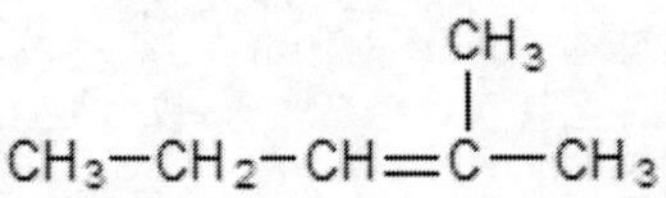 $CH_3-CH_2-CH=\underset{\underset{CH_3}{\mid}}{C}-CH_3$		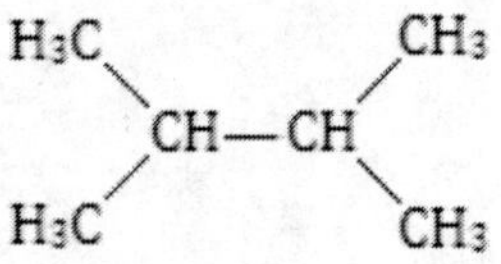 $\underset{H_3C}{\overset{H_3C}{\diagdown}}CH-CH\underset{\diagdown CH_3}{\overset{\diagup CH_3}{}}$	
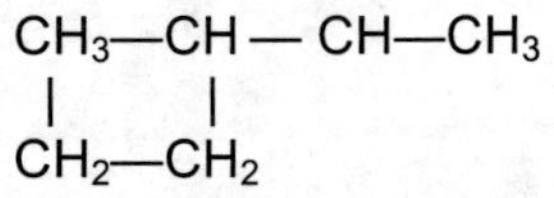 $CH_3-\underset{\underset{CH_2-CH_2}{\mid}}{CH}-\underset{}{CH}-CH_3$		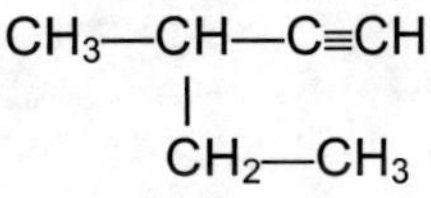 $CH_3-\underset{\underset{CH_2-CH_3}{\mid}}{CH}-C\equiv CH$	

3. Identify the following compounds as a(n) alcohol, ether, amine, aldehyde, ketone, carboxylic acid, ester, amide, or aromatic compound:

Molecule	Family	Molecule	Family
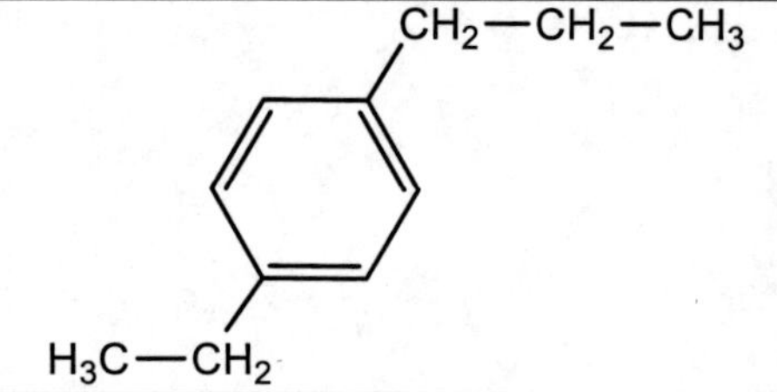 (benzene ring with CH_2—CH_2—CH_3 and H_3C—CH_2 substituents)		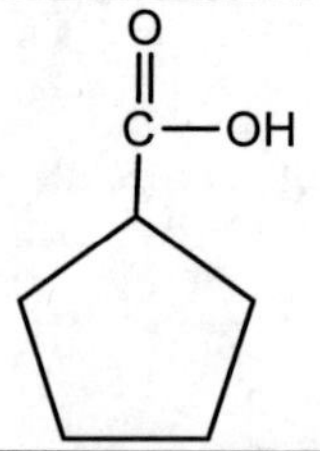 (cyclopentane ring with C(=O)—OH)	
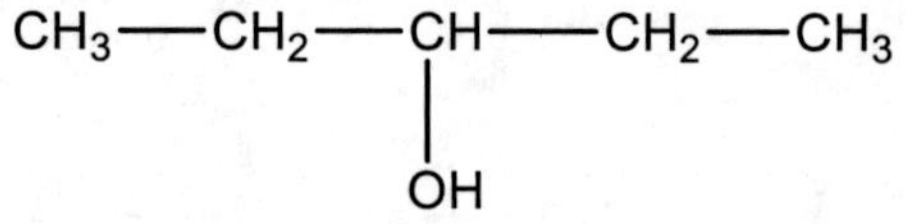 CH_3—CH_2—CH(OH)—CH_2—CH_3		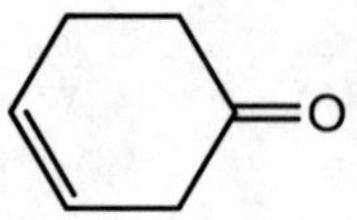 (cyclohexene ring with =O)	
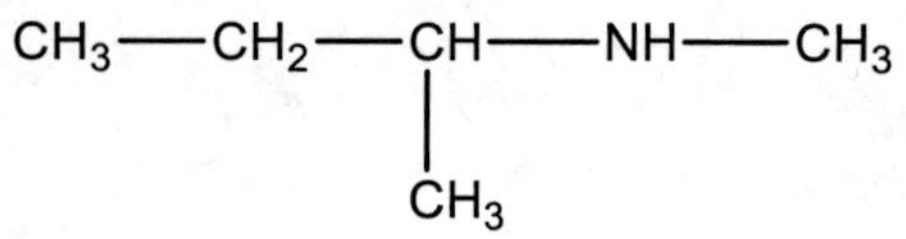 CH_3—CH_2—CH(CH_3)—NH—CH_3		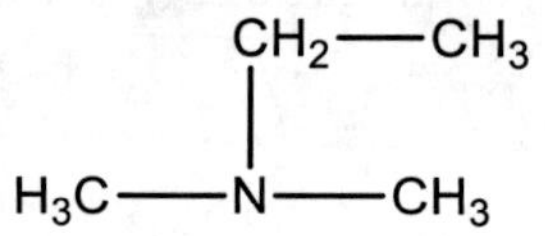 H_3C—N(CH_2—CH_3)—CH_3	
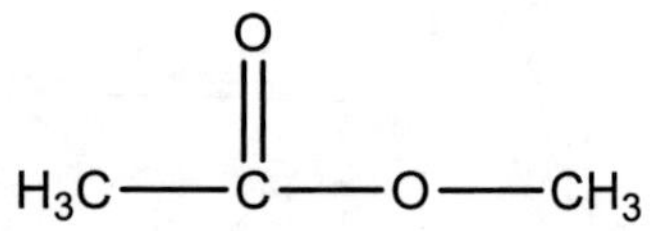 H_3C—C(=O)—O—CH_3		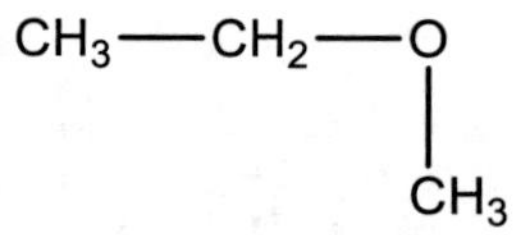 CH_3—CH_2—O—CH_3	
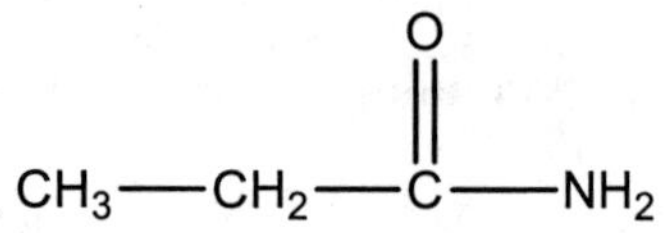 CH_3—CH_2—C(=O)—NH_2		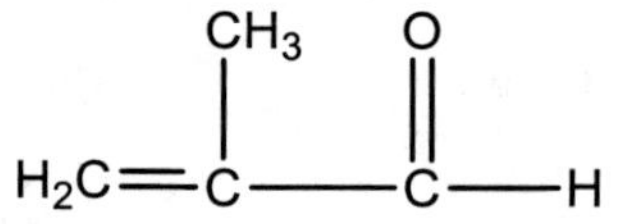 H_2C=C(CH_3)—C(=O)—H	

(DO NOT TURN IN THIS SHEET!)

In-Class Examples

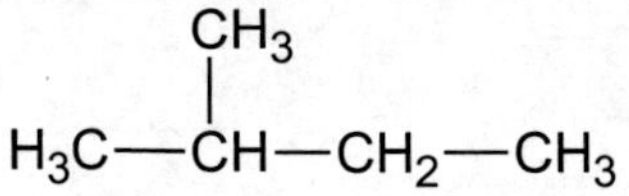

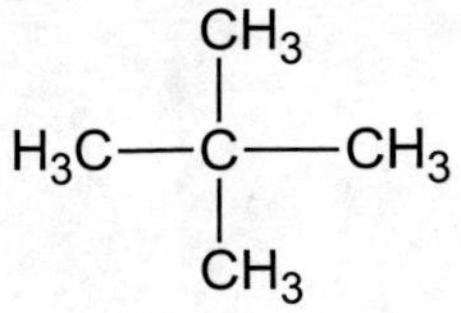

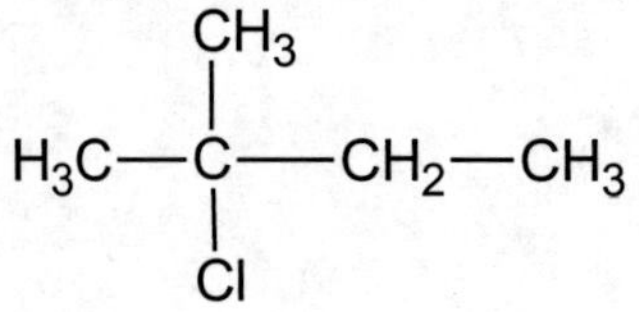

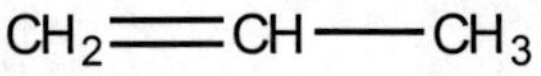

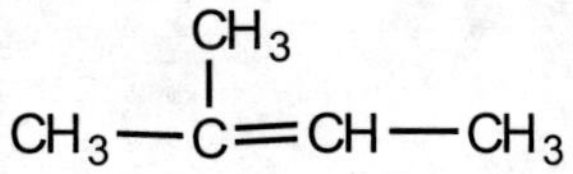

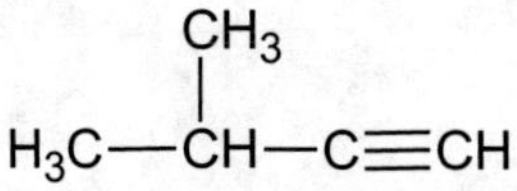

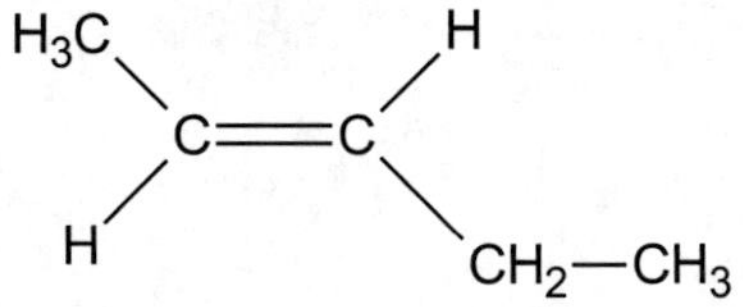

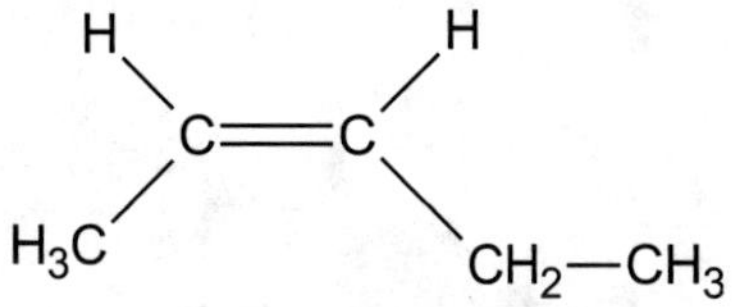

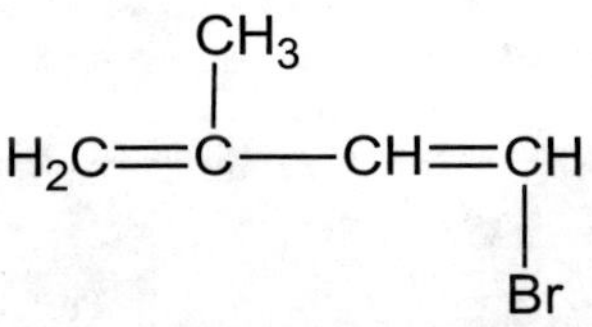

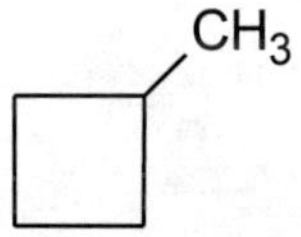

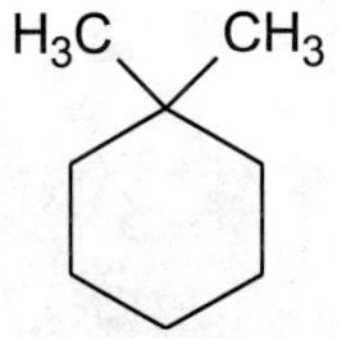

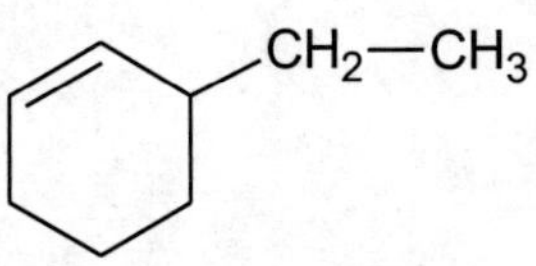

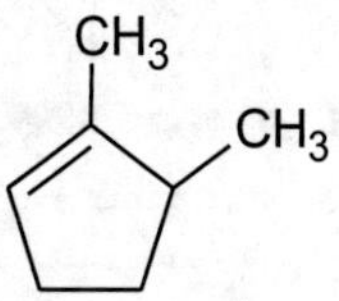

$H_3C-CH(CH_3)-CH_2-CH_3$

$H_3C-C(CH_3)_2-CH_3$

$H_3C-C(CH_3)(Cl)-CH_2-CH_3$

$CH_2=CH-CH_3$

$CH_3-C(CH_3)=CH-CH_3$

$H_3C-CH(CH_3)-C{\equiv}CH$

$H_2C=C(CH_3)-CH=CH(Br)$

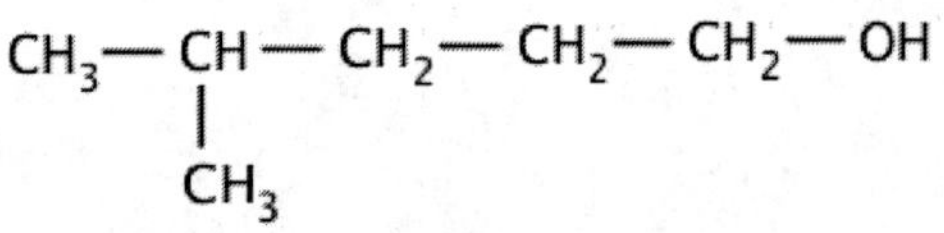

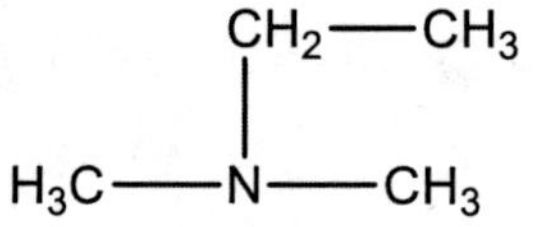

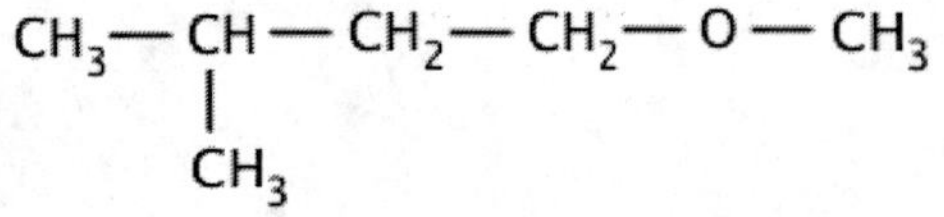

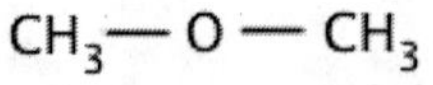

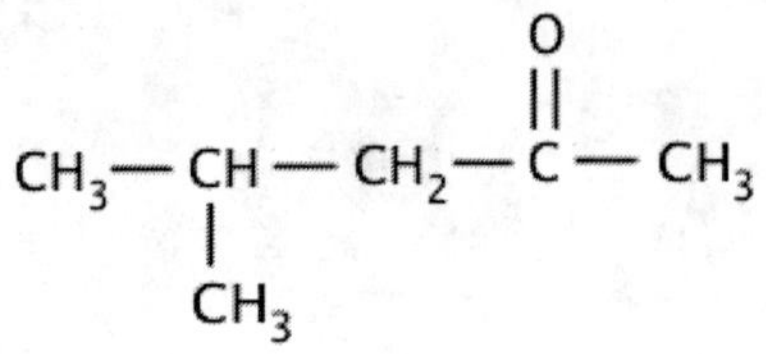

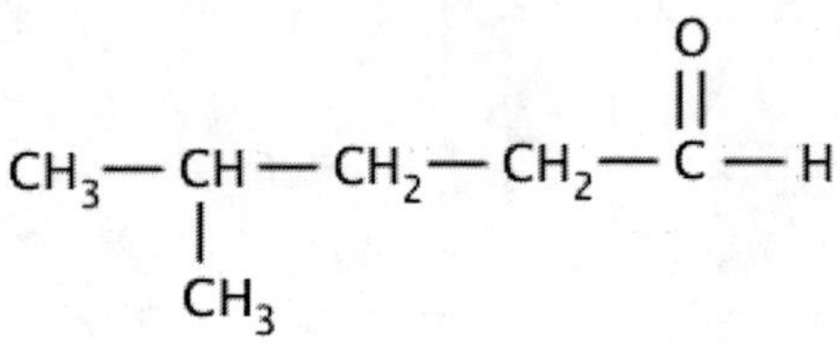

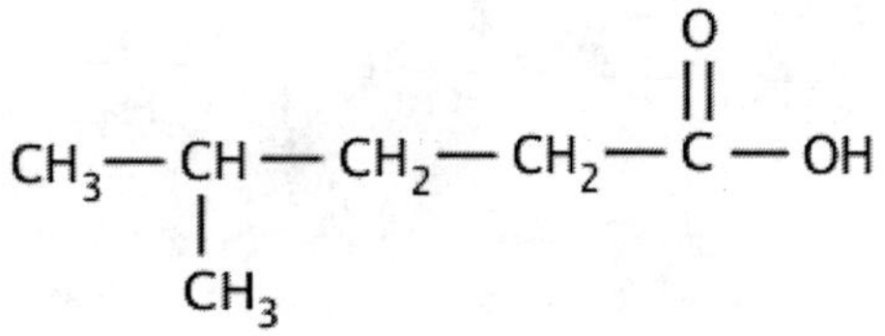

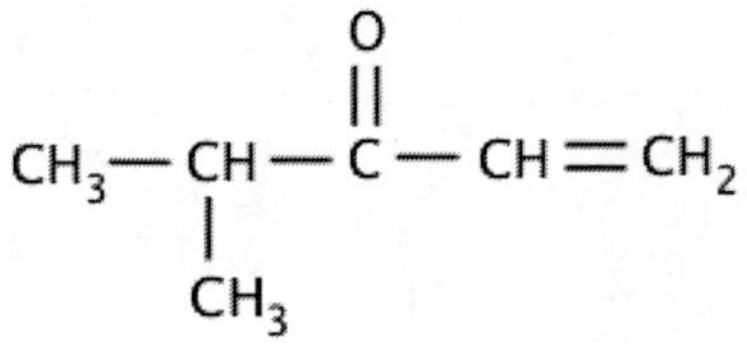

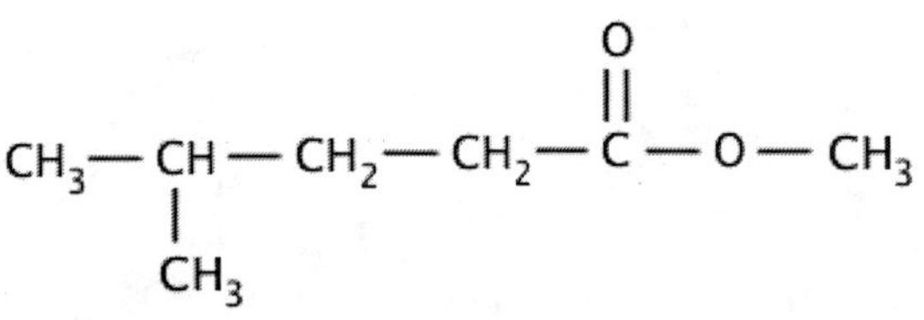

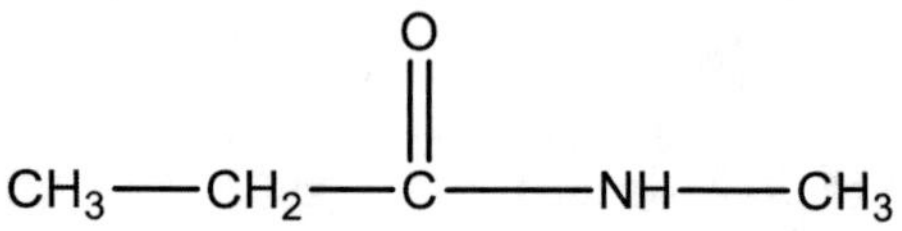

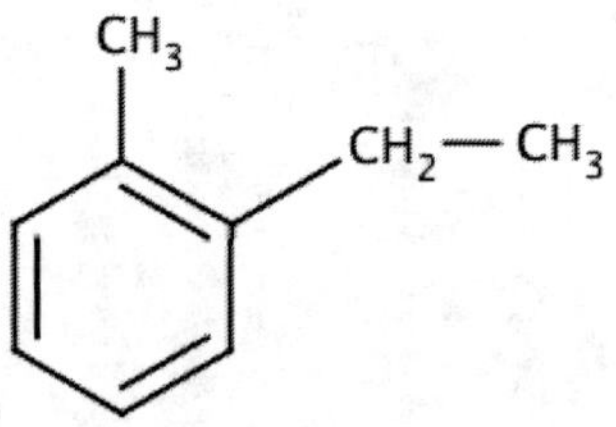

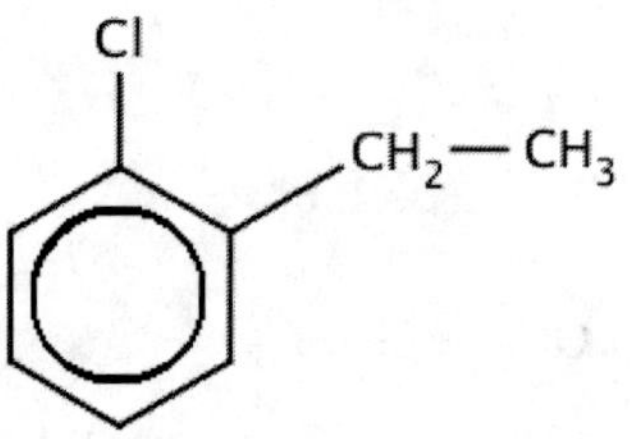

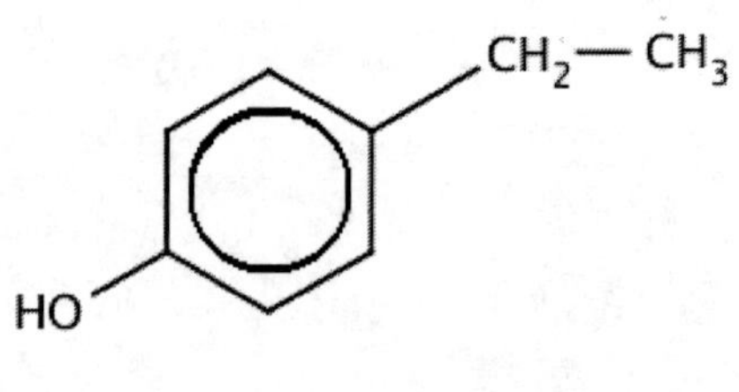

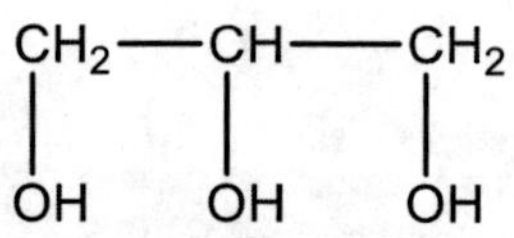

$CH_3-CH(CH_3)-CH_2-CH_2-CH_2-OH$

$H_3C-N(CH_2CH_3)-CH_3$

$CH_3-CH(CH_3)-CH_2-CH_2-O-CH_3$

CH_3-O-CH_3

$CH_3-CH(CH_3)-CH_2-C(=O)-CH_3$

$CH_3-CH(CH_3)-CH_2-CH_2-C(=O)-H$

$CH_3-CH(CH_3)-CH_2-CH_2-C(=O)-OH$

$CH_3-CH(CH_3)-C(=O)-CH=CH_2$

$CH_3-CH(CH_3)-CH_2-CH_2-C(=O)-O-CH_3$

$CH_3-CH_2-C(=O)-NH-CH_3$

$CH_2-CH-CH_2$ with OH, OH, OH

| name | | section | date |

Worksheet

1. Name the following hydrocarbons and halogenated hydrocarbons:

CH_3——CH_3

CH_3——CH_2——CH_3

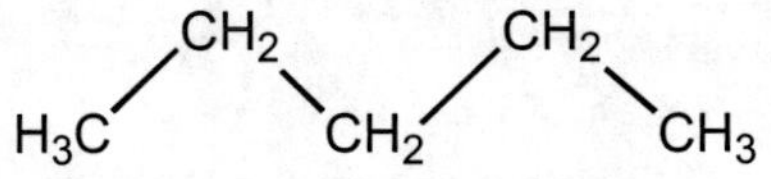

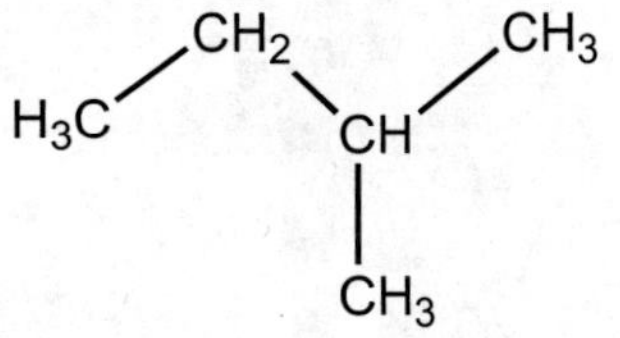

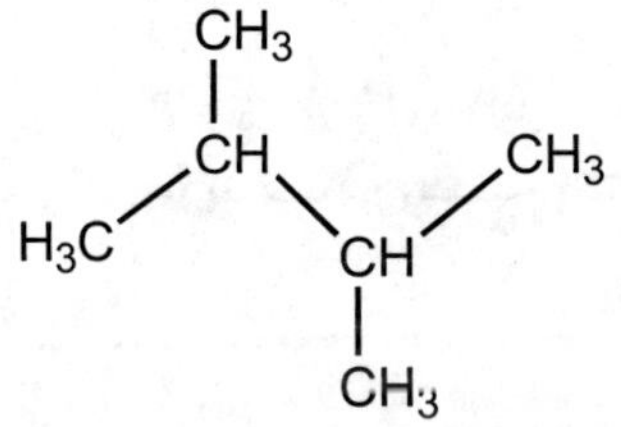

CH_3—$(CH_2)_5$—CH_3

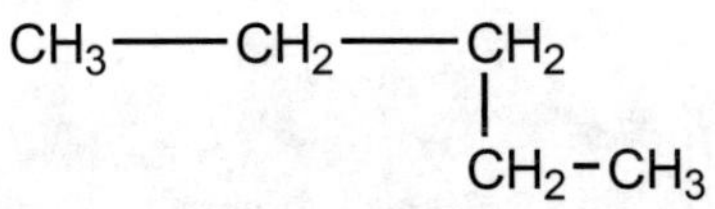

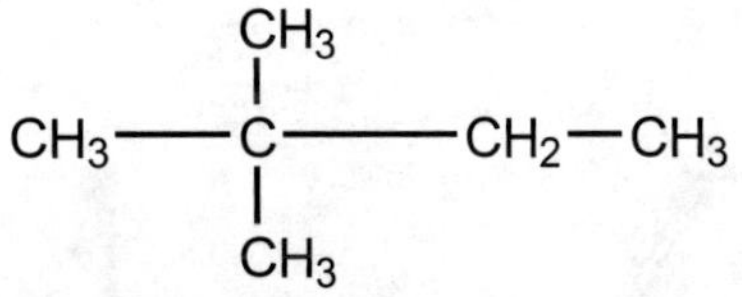

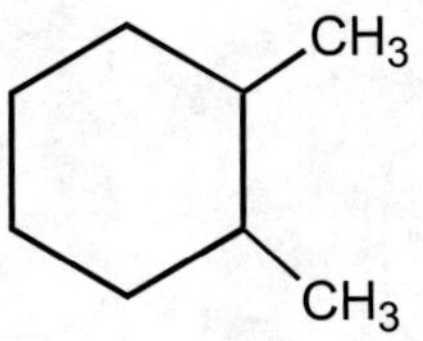

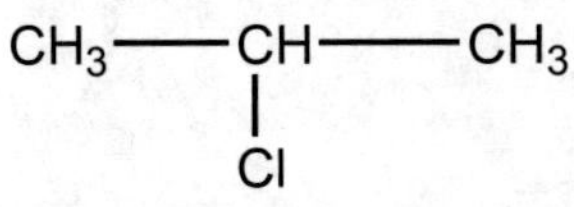

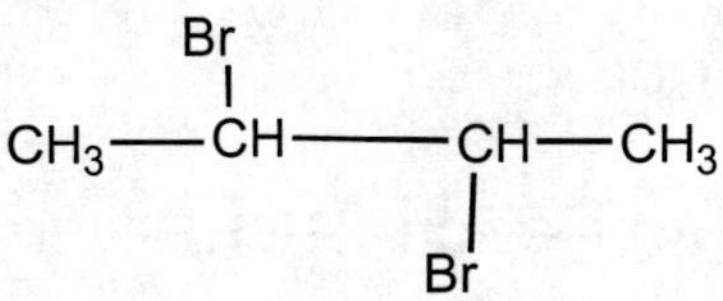

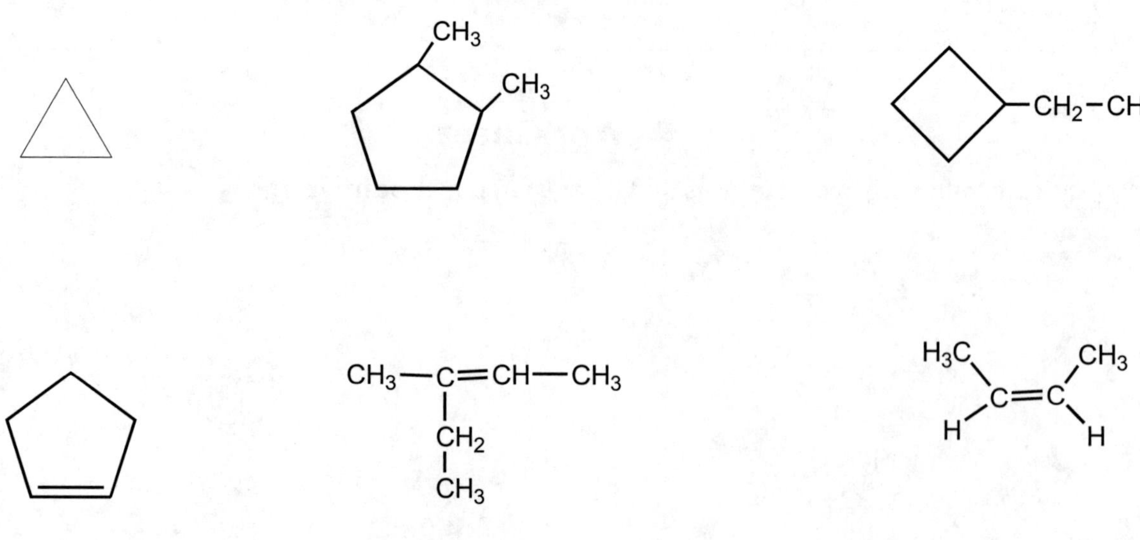

2. Name the following compounds with functional groups:

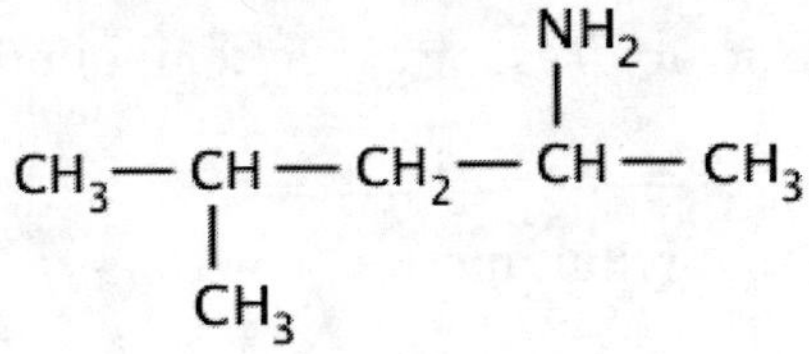

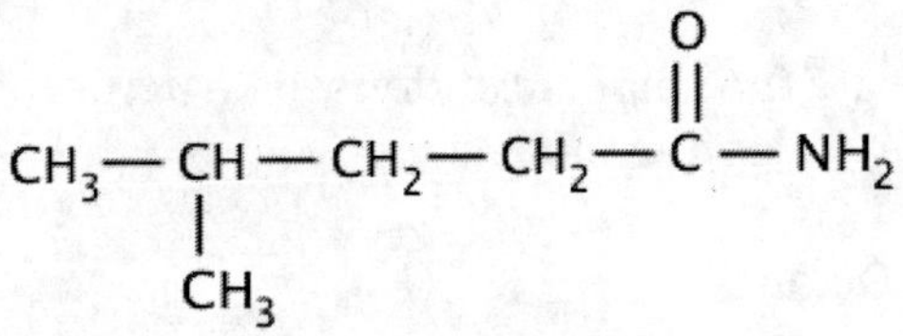

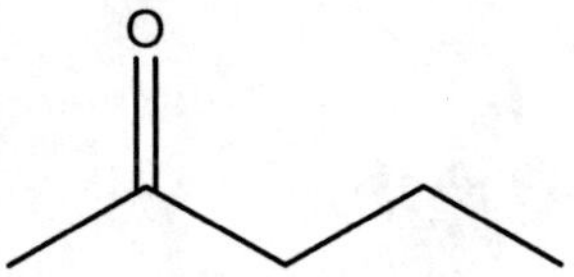

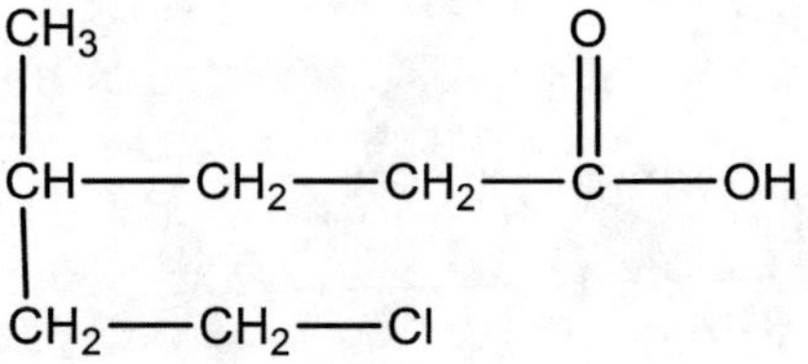

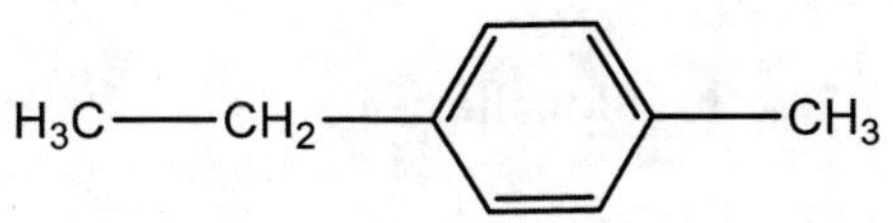

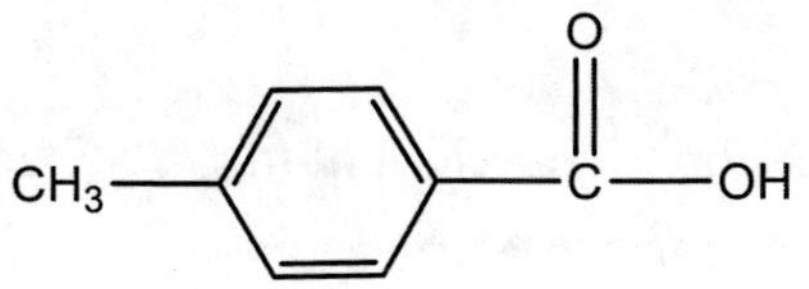

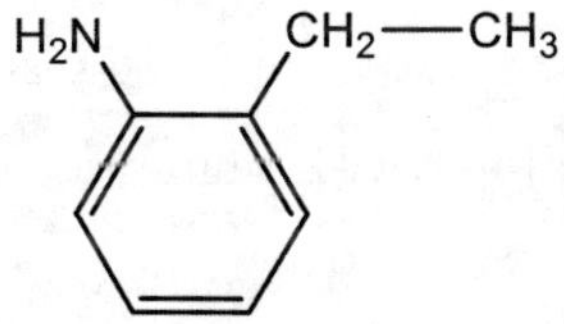

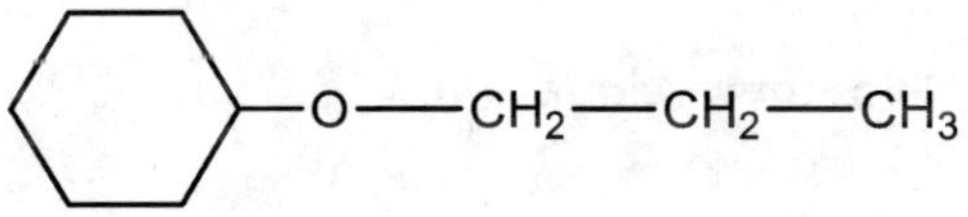

3. The following names are incorrect; correct them:

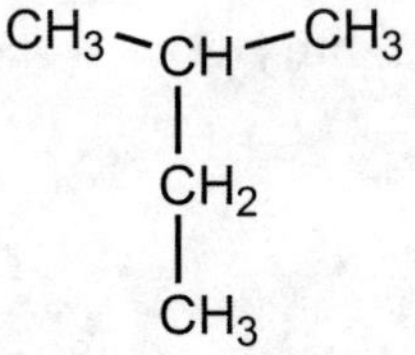

2-ethylpropane

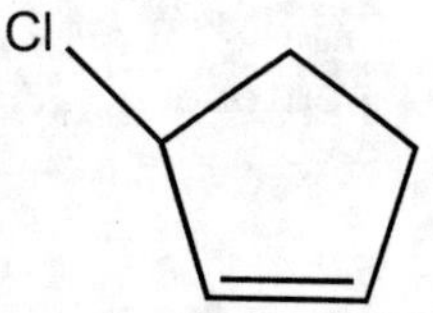

5-chlorocyclopentene

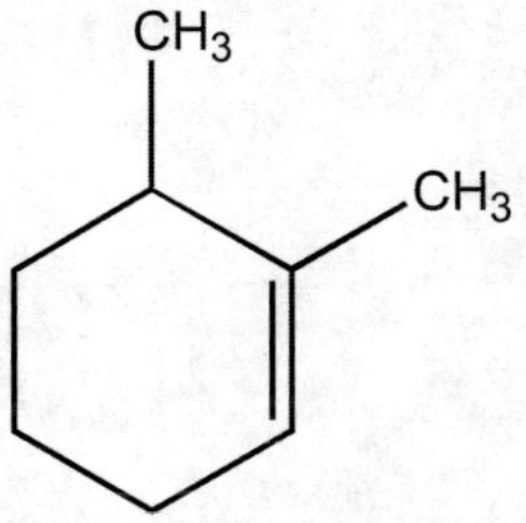

1,2-dimethylcyclohexene

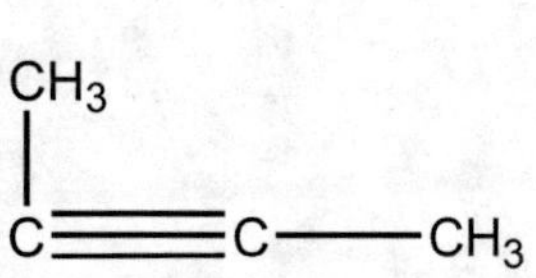

1-methylpropyne

4. Draw the correct condensed structural formulas (or line structures for cyclic compounds) for the following names:

Octane   2,3-dimethylbutane

2,2,5-trimethylhexane   2-bromo-3-methylbutane

1,3-dimethylcyclopentane   1,1,2,2-tetrachlorohexane

4-methylcyclohexene   *trans*-2,3-dimethyl-2-pentene

3,4-dimethyl-2-pentene   3-methyl-1-butyne

1,3-cyclopentadiene   ethyl propanoate

3,5-dichloro-2-hexanone

N,N-dimethylpropanamide

1-iodo-2-propanol

1,3-butanediol

5. **EXTRA CREDIT:** Draw condensed structural formulas of the five structural isomers of C_6H_{14} and provide IUPAC names for each. (*HINT*: There is one hexane, two pentanes, and two butanes. There are no ring structures among them.)

<table>
<tr><td>Isomer 1 structure:

Isomer 1 name:</td><td>Isomer 2 structure:

Isomer 2 name:</td></tr>
<tr><td>Isomer 3 structure:

Isomer 3 name:</td><td>Isomer 4 structure:

Isomer 4 name:</td></tr>
</table>

Isomer 5 structure:

Isomer 5 name:

EXPERIMENT 10: Polarity and Solubility

Michael Korn, Mike Goldin (Liberty University)

[T]hey will not adhere to one another, just as iron does not mix with clay.
– Daniel 2:43 (NKJV)

Background

Why does oil not mix with water or vinegar, but water and vinegar mix well? And what about so-called "water-soluble" and "fat-soluble" vitamins – what is the difference? The "secret" to knowing which two substances will or will not mix lies in their polarity! We have already learned how to predict whether small molecules will be polar or nonpolar and, from that, to predict the type of intermolecular forces that exist between the molecules of that pure substance. As it turns out, the same reasoning based on polarity applies to predicting whether two substances will mix.

For example, two liquids will mix together if both have dipole-dipole attraction forces (or hydrogen bonding forces – a special case of dipole-dipole) because both molecules are polar. If an ionic solid is being dissolved in a polar liquid, the similarity between the partial charges on the polar liquid molecules and the full-fledged charges on the ions of the solid, cause the attraction force between them, called **ion-dipole** attraction, to be similar to dipole-dipole as well, resulting in the ionic solid mixing, or being **soluble** in the liquid. (More details about how ionic compounds dissolve in water are given in Exp. 4 and 6). The same can be said about two nonpolar liquids – both would have London dispersion forces and would also mix well; such liquids are called **miscible** in each other.

The general idea of **solubility**, or which substances **dissolve** in which, can be stated as **like dissolves like** – with the "like-ness" of two substances being based on their polarity. In other words, polar substances will dissolve other polar substances (or the very similar ionic solids). Nonpolar substances will dissolve other nonpolar substances. However, the opposite is also true – a polar or ionic substance will not mix with a nonpolar substance! Two liquids that do not mix are called **immiscible** in each other; a solid or gas that will not mix with a liquid is said to be **insoluble** in it. (As a side note, the term **miscibility**, as you may have guessed, only applies to liquids, whereas the more general term **solubility** can be used with solids, liquids, and gases alike!) An example of miscible and immiscible liquids is shown in Figure 1. Please note especially how the miscible liquids form a solution (or a homogeneous mixture), so that it is impossible to tell two substances are present; at the same time, two immiscible liquids remain in two separate layers!

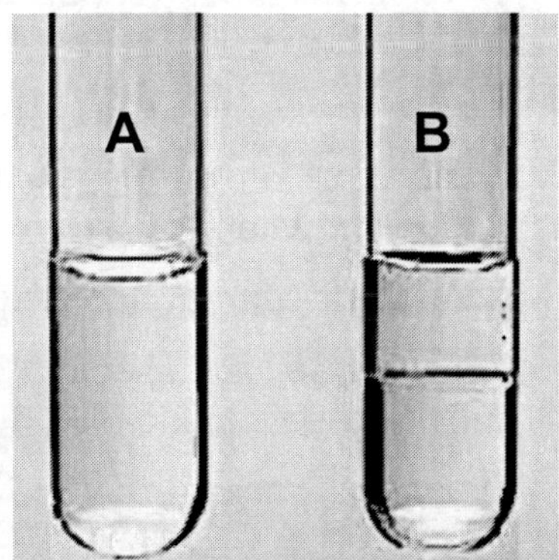

Figure 1. (A) Two miscible liquids; (B) two immiscible liquids

Of course, if we were to delve even deeper into polarity, we would find that it can be expressed quantitatively in terms of a number (called **dipole moment**), and we could reach more accurate conclusions about this "like-ness" of substances by comparing their dipole moments. However, if we at least have some idea of the polarity of two substances, we can still make a pretty good educated guess about their solubility in one another! This is quite easy for small molecules, for which we can easily determine whether they are polar or nonpolar based on the VSEPR shape of their molecules and the electronegativity difference between atoms in each of the bonds. But things get much more complicated when it comes to large molecules, especially of organic compounds.

The main reason for these complications for organic compounds is the fact that C—H bonds, with an electronegativity difference of roughly 0.4 on the Pauling scale, actually fall in the nonpolar category. For **hydrocarbons**, which are made up of strictly carbons and hydrogens, this makes it easy to determine their polarity – since there are no polar bonds, all hydrocarbons are "automatically" nonpolar! However, the polarity of organic molecules with so-called **functional groups**, or groupings of atoms containing an oxygen, nitrogen, etc., is much more difficult to simply predict – mainly because of the presence of polar C–O, O–H, C–N, N–H, or other bonds alongside a nonpolar "backbone" of carbons and hydrogens. In fact, this is precisely the issue that we will explore in this experiment.

We will try to predict the polarity of molecules containing both polar functional groups, which can be called "polar sites" and nonpolar carbon chains. Unlike small molecules, it is not possible to predict the polarity of the entire molecule simply from the 3-D shape around the central atom in the functional group. Generally, if the size (length) of the carbon chain is relatively small and not greatly larger than the size of the polar site, the molecule will be polar. If the nonpolar part is much larger, however, the molecule will be nonpolar. Without *knowing* the dipole moments of a specific molecule (or doing some lab experiments), we are only guessing, of course.

In this experiment, you will begin by making some guesses about the polarity of the compounds. From your predictions, you will then determine whether each molecule will be soluble in water and in cyclohexane. (Both water and cyclohexane are very straightforward in terms of their polarity.) You will then test your predictions by taking small samples of the actual liquids and mixing them with both water and cyclohexane.

Objectives

1. To predict the polarity of organic molecules based on their structure
2. To predict the solubility of organic molecules in polar and nonpolar solvents
3. To experimentally test the solubility of organic compounds in polar and nonpolar solvents

Safety

- Wear approved **safety glasses or goggles** during the entire experiment! It is especially important when we are working with toxic, corrosive, and irritant substances.
- All organic compounds used here are flammable!
- Concentrated acetic acid is corrosive and can irritate or cause burns, and several other substances are toxic. Avoid contact with skin! The use of gloves is highly recommended.
- Many of the organic compounds used here are volatile and must be used under a fume hood.
- Remember the location of the emergency eyewash and safety shower in case of accidents!
- Wash your hands thoroughly before leaving the laboratory.

Materials and Equipment

You will need the following for your experiment:

1. Small test tubes (6)	6. Ethanol	11. Ethyl acetate
2. Glass rod	7. Isopropanol	12. Wintergreen oil
3. Molecular model kit	8. Butanol	13. Xylene
4. Tap water	9. Acetone	14. Sodium chloride
5. Cyclohexane	10. Acetic acid	15. Sodium acetate

Procedure

I. **Polarity and Solubility Predictions**

Record all predictions in the appropriate row and column of the table on the Worksheet before you proceed with the solubility experiments in Part II.

1. For each molecule in the table on the Worksheet, identify the polar and non-polar parts in the molecule. In each molecule, circle the nonpolar part(s) and box the functional groups that are polar sites.

2. Predict the overall polarity of the molecule based on the relative size of polar sites and nonpolar parts of each molecule. **You will not be penalized for getting this part wrong— this is just an educated guess based on your best understanding of polarity to this point!**

3. Based on the predicted polarity of each molecule from Step 2, predict the solubility of each compound in (or miscibility with) water and cyclohexane.

4. Have the instructor or TA initial your predictions for completion.

II. **Solubility Experiments**

Record all data in the appropriate row and column of the table on the Data/Observations sheet.

> **_IMPORTANT:_** *Please use the pipettes (eye droppers) provided with each solution bottle for this purpose! Do not use the same pipette for different solutions—this will lead to **contamination** and errors in everyone's observations!*

5. Take one test tube and fill it with approximately 1 mL tap water using a plastic pipette. To the same test tube, add approximately 1 mL cyclohexane using the plastic pipette provided with the dispensing bottle. Observe whether the liquids are miscible and record your observations.

6. Take five test tubes and fill each with approximately 1 mL of cyclohexane using the plastic pipette provided with its dispensing bottle. Using the plastic pipettes provided with each solution, to each add approximately 1 mL of liquids numbered 1-5 on the Data/Observations sheet.

7. **Mix the contents well, either by shaking and swirling or by using a glass rod.** Observe whether the liquids are miscible and record your observations. Record a "Yes" if the two substances mixed together and formed a solution (homogeneous mixture); record a "No" if the two substances did not mix.

    (*HINT: Note carefully which test tubes contain which mixtures. Alternatively, use labeling tape and a permanent marker to label each test tube.*)

8. Discard the mixtures into the collection container labeled "MIXED ORGANIC SOLVENTS," taking care not to get them on your hands.

9. Once again fill each with approximately 1 mL of cyclohexane using the plastic pipette provided with its dispensing bottle. Using the plastic pipettes provided with each solution, to each add approximately 1 mL of liquids numbered 6-8—or, in the case of solids numbered 9-10, a small amount on the tip of a spatula.

10. Repeat steps 6-7 with this group of samples and record your observations.

11. Wash the five test tubes with soap and water. Then fill each with approximately 1 mL of tap water using a plastic pipette. Using the plastic pipettes provided with each solution, to each add approximately 1 mL of liquids numbered 1-5 on the Data/Observations sheet.

12. Repeat steps 6-7 with this group of samples and record your observations.

13. Wash the five test tubes with soap and water. Then once again fill each one with approx.imately 1 mL of tap water using a plastic pipette. Using the plastic pipettes provided with each solution, to each add approximately 1 mL of liquids numbered 6-8—or, in the case of solids numbered 9-10, a small amount on the tip of a spatula.

14. Repeat steps 6-7 with this group of samples and record your observations.

15. At this time also discard the water/cyclohexane mixture from the test tube in Step 1 into the collection container labeled "MIXED ORGANIC SOLVENTS," taking care not to get it on your hands.

Analysis/Conclusions

1. For each of the substances labeled 1-10, use your data on solubility/miscibility, as well as what you know about the polarity of water and cyclohexane, to determine the polarity of each one. Record your conclusions about each substance in the column titled "Experimental Polarity" on the Data/Observations sheet.

2. Under "Conclusions" on the Data/Observations sheet, answer questions 1 and 2:

    o For Question 1, compare your predictions of solubility/miscibility to the experimental data. Did they match up? If so, what made your prediction successful? If not, at what point in your prediction did you make a mistake? (*HINT: Consider how you arrived at the solubility/miscibility predictions!*)

    o For Question 2, another way of stating it is this: For all the compounds in the experiment that were **soluble** in water, were they also **insoluble** in cyclohexane? And vice versa, were all the compounds **insoluble** in water also **soluble** in cyclohexane? Whether all of them matched this way or not, how do you explain it?

Clean-Up

- Clean up any spills with soap and water. Do this as soon as you find a spill!

- Discard the contents of each test tube into the collection container labeled "MIXED ORGANIC SOLVENTS," taking care not to get them on your hands. Do this as you are following the directions in Procedure (you will have four batches of five test tubes to discard in addition to the water/cyclohexane test tube from Step 1).

- Wash all of the test tubes with soap and water using the test tube brushes located by each of the two large sinks.

______________________________   __________   __________
name   section   date

Pre-Laboratory Assignment

1. Define the terms below.

    a. Immiscible –

    b. Hydrocarbon –

    c. Functional group –

2. What is the general rule of solubility? *Explain briefly what it means!*

3. For a molecule of phosphorus trichloride, PCl_3:

    a. Draw the Lewis structure of this molecule. Using the Pauling electronegativity scale, determine which bonds are polar; show them on the structure using the $\delta+/\delta-$ notation.

    b. Is this molecule polar? *Explain briefly why or why not.*

    c. What type of intermolecular forces should exist between two PCl_3 molecules?

    d. Will PCl_3 be soluble in water, H_2O? *Explain briefly why or why not.*

4. On the Worksheet, add any missing lone pairs. To avoid getting points off, have the TA or instructor check this step for completion before you turn in the pre-lab.

____   __________   __________
name   section   date

Worksheet

I. Predictions of Polarity and Solubility

Name	Lewis Structure	Polarity	Soluble in C_6H_{12}?	Soluble in H_2O?
Water	H–O–H			✕
Cyclohexane (C_6H_{12})	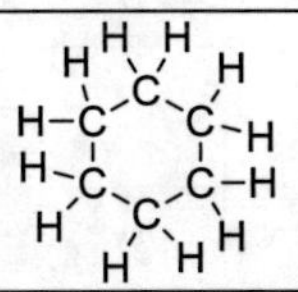		✕	
1. Ethanol	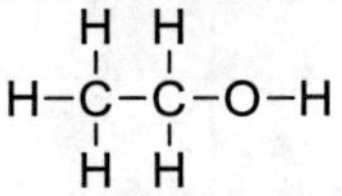			
2. Isopropanol	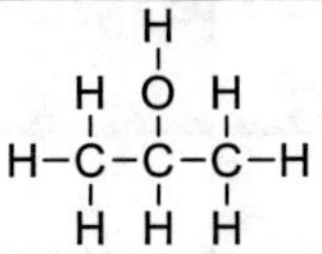			
3. Butanol	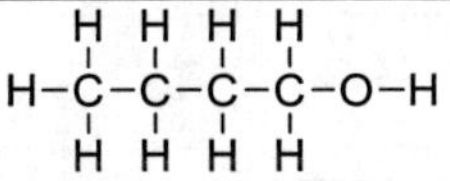			
4. Acetone	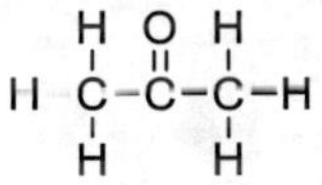			
5. Acetic acid	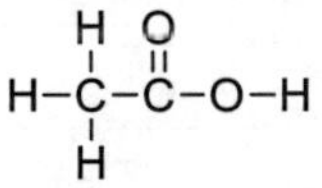			
6. Ethyl acetate	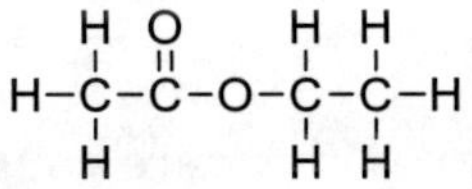			
7. Wintergreen	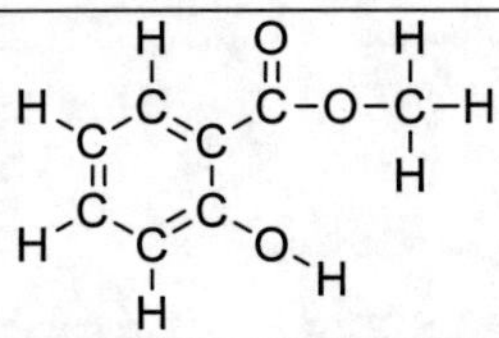			
8. Xylene	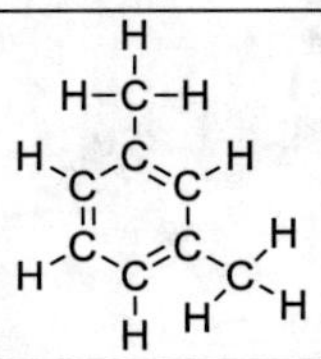			
9. Sodium chloride	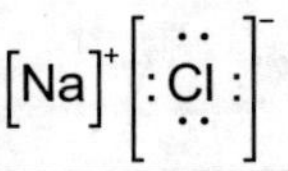			
10. Sodium acetate	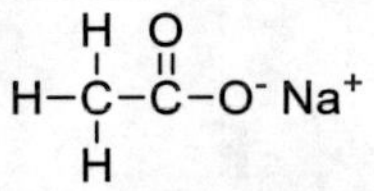			

Data/Observations

II. Solubility Experiments

Name	Soluble in C_6H_{12}?	Soluble in H_2O?	Experimental Polarity
Water		✕	
Cyclohexane (C_6H_{12})	✕		
1. Ethanol			
2. Isopropanol			
3. Butanol			
4. Acetone			
5. Acetic acid			
6. Ethyl acetate			
7. Wintergreen			
8. Xylene			
9. Sodium chloride			
10. Sodium acetate			

Conclusions

1. Which of the compounds did not follow your prediction? Why were your predictions wrong?

2. Do all the "Yes"es in the "Soluble in H_2O" column above also have a "No" for the "Soluble in C_6H_{12}" column? Why is this?

name   section   date

Post-Laboratory Assignment

1. Examine the structural formulas of the three biological molecules A, B, and C shown below.
    (*NOTE: The zigzag lines and polygons represent carbons with C–H bonds omitted!*)

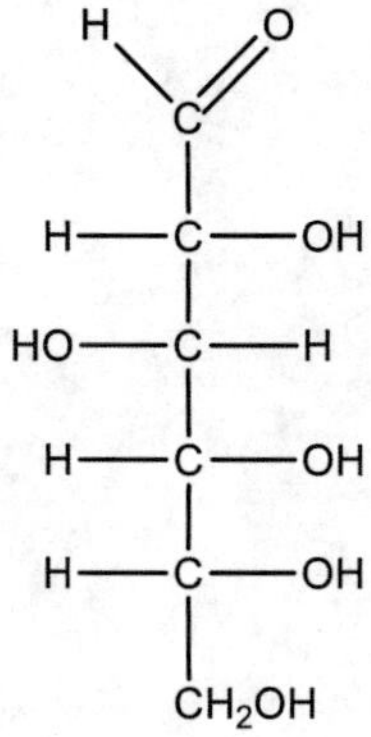

A 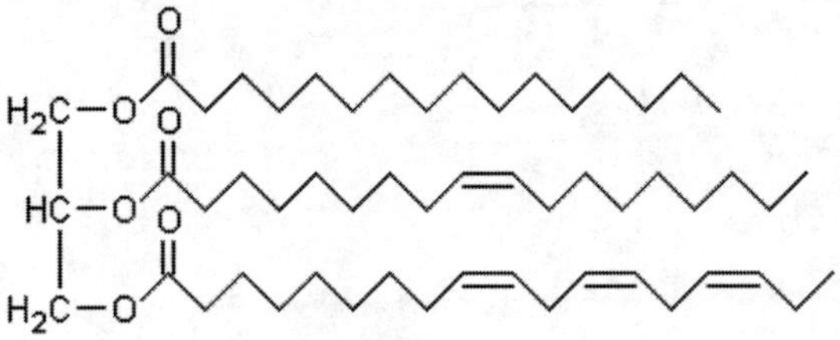   B 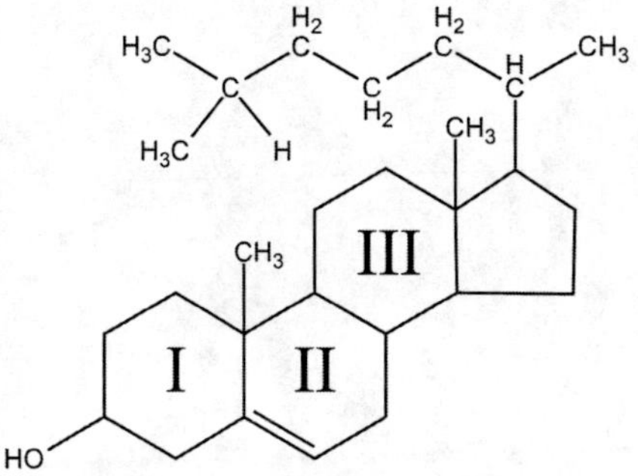   C

a. Circle the polar sites on each of these molecules.

b. Which of these molecules is/are soluble in water? *Explain briefly.*

2. Based on structures, predict the polarity of the following vitamins. Also, predict whether they
    are water-soluble or fat-soluble. (*HINT: Fat is nonpolar!*)

	Vitamin A 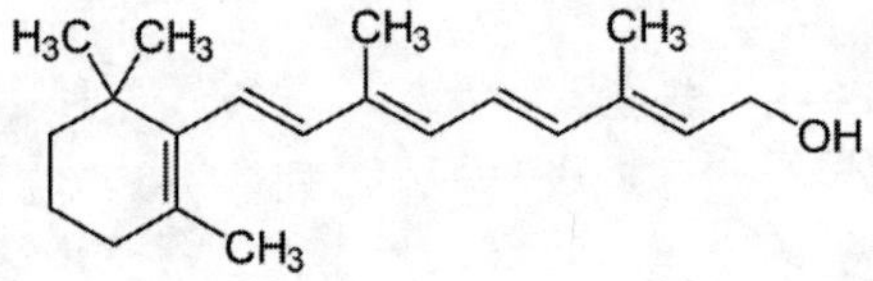	Vitamin C 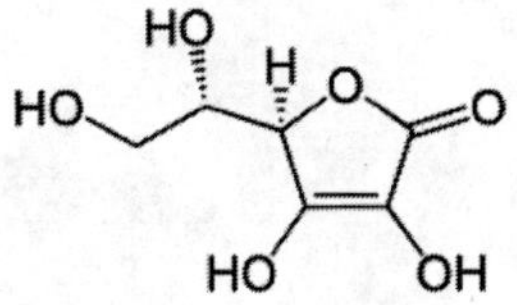	Vitamin D 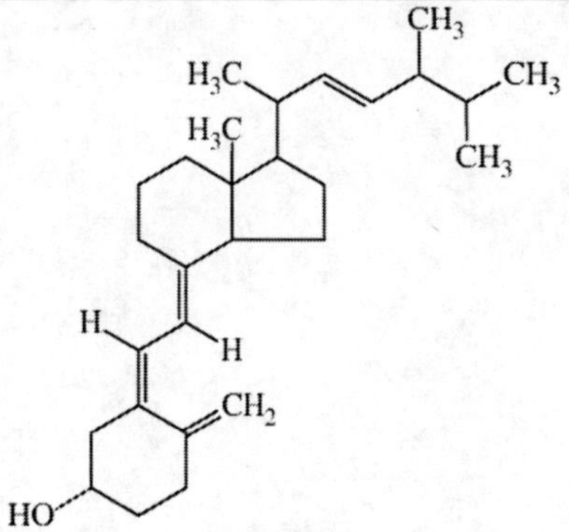
Polarity:			
Solubility:			

Mark Hemric, Mike Goldin (Liberty University)

> **_All the nations will be gathered before [Christ], and He will <u>separate</u> them one from another,_**
> **_as a shepherd divides his sheep from the goats._**
>
> – _Matthew 25:32 (NKJV)_

Background

Mixtures are very common in our lives: coffee, soda pop, chicken noodle soup, gasoline, and even the air we breathe are all "everyday" examples of mixtures. In fact, we come across very few materials that are **pure substances**, not mixed with other substances. Anything that contains two or more substances that are not chemically combined is a **mixture**.

It is easy to tell that the **texture** of some mixtures appears non-uniform—that that they consist of two or more components, like chicken noodle soup or muddy water; these are called **heterogeneous**. Others appear as if there was only one substance; these mixtures are said to be _uniform_, or the same, throughout and are called **homogeneous**. For instance, there is no visible difference between water (a pure substance) and salt water (a mixture of two substances—table salt and water). We usually call homogeneous mixtures **solutions**. The component of a solution present in the greatest amount is called the **solvent**, and any other components are **solutes**.

The pure substances contained in any mixture can be divided from one another; each component can be _separated_ from the others and _purified_. Separating mixture components always involves some kind of **physical change**. This change must involve a **physical property** that is different for the components being separated. For example, one substance dissolves in water and the other doesn't, or one is attracted to a magnet and the other isn't, etc. Table 1 shows some common separation techniques. Usually, a few of these must be used together to fully separate a mixture.

<u>**Table 1.**</u> **Techniques for separation of mixtures**

Technique	Uses	Summary	Property Used
Decantation	Separating _heterogeneous_ solid-liquid mixtures	Pouring off a liquid without disturbing the solid	Liquids flow, but solids do not
Distillation	Separating _homogeneous_ liquid-liquid mixtures (solutions)	Heating the solution to evaporate one component, then condensing it separately	Liquids have different boiling points and will evaporate from a mixture
Evaporation	Separating _homogeneous_ and/or _heterogeneous_ mixtures	Evaporating the liquid component, while the solid is left behind	Liquids evaporate at temperatures lower than needed to melt or sublimate solids
Extraction	Separating various mixtures	Using a liquid **solvent** to dissolve (mix with) one component, leaving another component behind	Different substances have different abilities to dissolve in liquids
Filtration	Separating _heterogeneous_ solid-liquid mixtures	Using a porous material (a **filter**) to trap the solid (called **residue**) and allow the liquid (called **filtrate**) to pass through	Solids cannot pass through porous materials; liquids can
Sublimation	Removing one component of a solid-solid mixture	Heating a mixture of solids until one solid sublimates	Some solids will sublimate when heated (only works for solids that do this!)

Alcohol in Mouthwash and Beverages

Alcohols is a family of compounds belonging to a large class of substances classified as **organic compounds**. (We will learn much more about them in the weeks to come!) The word "alcohol" in this lab, however, is used in the everyday sense of "grain alcohol," or **ethanol**. Probably the best-known use of alcohol is in alcoholic beverages that are solutions of alcohol and water with minor amounts of other solutes present. The alcohol content in beer is usually from 3 to 8%, whereas wines are usually 8 to 14%. To achieve beverages with higher alcohol content, the alcohol must be **distilled** from beverages produced by fermentation. (Distillation is discussed below in detail.)

The history of ethanol being used as a "euphoric beverage" goes back into ancient history—after all, the account of Noah getting drunk in Genesis 9:20-27 takes place shortly after the Flood and is the first account of anyone in the Bible being drunk. The oft-quoted verse, Ephesians 5:18, contains a clear commandment: "do not be drunk," or as we would put it in the 21st century, "do not become intoxicated" with alcoholic beverages. While various Christian denominations have different positions on whether it is okay for a believer to consume alcoholic beverages "socially," there is no question what the pointed words of Ephesians 5:18 mean—being drunk is a direct violation of a Scriptural commandment!

Liberty University's position on alcohol consumption is a total prohibition of the possession, use, manufacture, or distribution of alcoholic beverages (and other controlled substances and illegal drugs) by LU students. This is a long-standing University policy, rooted in Liberty's spiritual heritage and concern for students' well-being. There is also a good medical rationale for this—ethanol is a mind-altering drug, much like narcotic drugs, causing a variety of symptoms from poor judgment to impaired thinking and bodily control all the way to unconsciousness and death. A significant number of car accident deaths occur in DUI crashes. Heavy alcohol drinkers also experience serious long-term illness because of permanent damage to the brain and other organs (such as cirrhosis of the liver, for example).

In spite of ethanol's more notorious use in beverages, its largest use is in motor fuel. Many gas stations in the U.S. have begun carrying gasoline with added ethanol. It is also used to kill germs and is a common solvent in chemistry. In mouthwash, contrary to what one might expect, the main use of ethanol is to help dissolve the active ingredients that do not easily dissolve in pure water. In this experiment, we will use a simple distillation apparatus to purify alcohol from mouthwash.

Distillation and Boiling Points

Distillation is a process of separating components of a liquid mixture based on differences in their boiling points by evaporating one of the components of the mixture and then condensing the released gas when it encounters colder surfaces. (The resulting liquid, called **distillate**, is collected.) Therefore, it is important to understand the basics of how a liquid (or a mixture) boils. For a pure liquid to boil, a temperature has to be reached when the pressure of the vapor formed when the liquid is evaporating is equal to the pressure of the atmosphere around it. For example, the vapor pressure of steam above the surface of water at 100°C is equal to 1 atm, which is the atmospheric pressure at sea level.

For mixtures, this gets a bit more complicated, as the content of the vapor above the liquid mixture depends on how easily each liquid evaporates, as well as on the amount of each liquid. If one component is much more likely to evaporate, most of the vapor formed above the liquid will contain only that substance. To put it differently, one of the liquids will boil first, and eventually all of it will evaporate and leave the other liquid behind. On the other hand, if two liquids evaporate equally well, both will be present *in the same proportion* in the vapor and in the liquid mixture. In this case, the whole mixture boils at the same temperature and cannot be separated by distillation.

Finally, if a liquid is almost pure, with only a tiny amount of another liquid that is more difficult to evaporate, its boiling point will be slightly higher than the boiling point of the totally pure liquid. The effect of the solute increasing the boiling point of a solvent is called **boiling point elevation** and is also observed with *solids* dissolved in a liquid. It is related to the vapor pressure of the mixture being lower because the solid (or small amount of liquid) is not evaporating and is considered **nonvolatile**.

Water and ethanol mix in all proportions—we say that they are **miscible**. The boiling points of ethanol (78.3°C) and water (100.0°C) are far enough apart to allow us to separate them by distilling the alcohol. Ethanol evaporates, is condensed by cooling, and is collected as the distillate. It is not possible, however, to completely separate ethanol and water because they begin to boil together once the amount of ethanol reaches about 96%. (While many students report results of 100%, this is a result of experimental error—most likely having impurities other than water in the recovered ethanol, such as the active ingredients of the mouthwash). We can determine the purity of the ethanol recovered in two ways.

The easiest way is to measure the density of the distillate and then compare it to known densities of water-ethanol mixtures. Since these mixtures are well known, we can use data published in literature to plot a graph (shown in Figure 2, p. 51) to determine the percentage of ethanol from its measured density. To calculate its density, we will measure the mass and volume of the recovered ethanol.

Another way to confirm the presence of small amounts of impurities in alcohol is to measure the boiling point of the distillate. Recall that the boiling point of *pure* ethanol is 78.3°C; if the distillate is a mixture that contains a small amount of water, you would expect it to have a boiling point a little above 78.3°C due to boiling point elevation. However, it is crucial to know what the solvent is: for example, if most of the alcohol is lost (for example, due to the ice bag falling off) and your distillate is mostly water with a little bit of alcohol or other solutes, then water is the *solvent,* so the boiling point of such a mixture will be above 100.0°C, not 78.3°C!

Objectives
1. Perform distillation in a simple distillation apparatus
2. Measure boiling points of liquids
3. Discover and explain the effect of a solute on a solvent's boiling point
4. Use the calculated density of a mixture along with a graph of density vs. concentration to determine the concentration of the solute

Safety
- Wear approved **safety glasses or goggles** during the entire experiment!

- Alcohol and active ingredients of the mouthwash can irritate eyes on contact.

- As always, avoid **burns** by contact with hot objects! Do not touch hot glassware or hot plate surfaces. Never hand hot objects to another person! Avoid contact with hot steam that may come out during distillation.

- Pasteur pipettes have thin glass tips that break off easily. Take care not to cut or stab yourself with it!

- Wash hands thoroughly before leaving the laboratory.

Materials and Equipment

You will need the following for your experiment:

1. Filtration flask, 250 mL
2. Hot plate
3. Beaker, 250 mL or 400 mL
4. Test tube, small
5. Graduated cylinder, 10 mL
6. Additional graduated cylinder, 10 mL *
7. Pasteur pipette with plastic tubing *
8. Rubber stopper with a thermometer *
9. Stir bar *
10. Plastic snack bag with crushed ice *
11. Rubber band *
12. Digital thermometer *

Procedure

1. Obtain the equipment marked in the list above with an asterisk "*" from the front desk.

2. Weigh two clean, dry 10-mL graduated cylinders; record their masses on the Data Sheet.

3. Obtain about 50 mL of mouthwash in a 250-mL filtration flask from your instructor. Place a stir bar in the flask.

4. Assemble the distillation apparatus shown in Figure 1:
   - Place the stopper with a thermometer in the mouth of the flask;

   - Place the glass Pasteur pipette with Tigon tubing on the side arm of the flask;

   - Half-fill a reclosable plastic bag with ice;

   - Wrap the ice bag around the pipette and fasten it with two rubber bands;

   - Place the assembly on a hot plate; place one of the 10-mL graduated cylinders that you weighed in Step 1 so that it catches the condensing alcohol from the pipette.

     (You may have to place something underneath the graduated cylinder to bring it closer to the pipette; a piece of Styrofoam usually works well.)

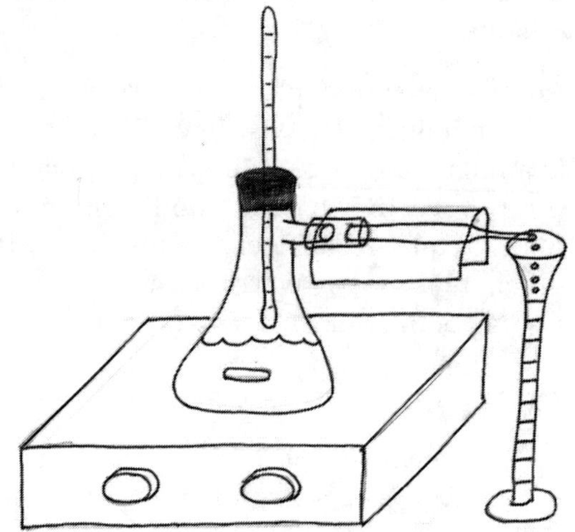

Figure 1. Distillation apparatus

5. On the hot plate, turn the "STIR" knob to 150 rpm and the "HEAT" knob to 230°C.

6. After 15-20 minutes, you should reach about 80-85°C on the thermometer and begin seeing bubbles at the very surface of the mouthwash, as well as vapor condensing on the flask walls. After a few more minutes, distillate liquid should begin dripping from the Pasteur pipette into your 10 mL graduated cylinder. Collect 3-5 mL of the distillate into the graduated cylinder.

7. Remove the first graduated cylinder and begin collecting distillate in another graduated cylinder. Measure the volume of distillate and the mass of the graduated cylinder with the distillate. Record both of these on the Data Sheet.

8. Collect 3-5 mL of the distillate into the second graduated cylinder. Measure the volume of distillate and the mass of the graduated cylinder with the distillate; record it on the Data Sheet.

9. Turn the "STIR" knob on the hot plate until the stirrer is off and take the distillation apparatus off the hot plate. Turn up the "HEAT" knob to a setting of 300-325°C and place a beaker of warm tap water on the hot plate. This is the hot water bath you will use in the steps below to determine the boiling point of the recovered alcohol. Then proceed to Steps 10-11 quickly, before the water starts boiling.

10. Transfer the contents of each graduated cylinder into a separate test tube and place a small boiling stone into each one.

    (You can use your graduated cylinder instead of transferring to a test tube if your cylinder has a removable plastic stand on the bottom; this needs to be removed prior to the next step and replaced at the end! You also need to place a boiling stone inside the cylinder.)

11. Place the test tubes (or graduated cylinders) into the hot water bath. When the distillate begins to boil, place a digital thermometer into the test tube and record its boiling point.

Clean-Up

- Discard the ethanol and the boiling stone into the container labeled "DISCARD ETHANOL."
- Use a magnetic catcher to "fish out" the stir bar from the remainder of the mouthwash in the filtration flask.
- Pour the remaining liquid into the container labeled "DISCARD MOUTHWASH."
- Wash any glassware; clean up any messes **_with soap_** (dried mouthwash is very sticky!!).
- Return borrowed equipment (marked with "*" in the list above) to the front desk.

Calculations

1. Calculate the mass of the distillate for each trial by subtracting the mass of the graduated cylinder alone from the combined mass of the graduated cylinder and distillate:

$$m_{\text{distillate}} = m_{\text{distillate + container}} - m_{\text{container}} \qquad \text{[1]}$$

2. Calculate the density of the distillate for each trial and use Figure 2 on the next page to match the percentage of ethanol to calculated density of distillate for each trial.

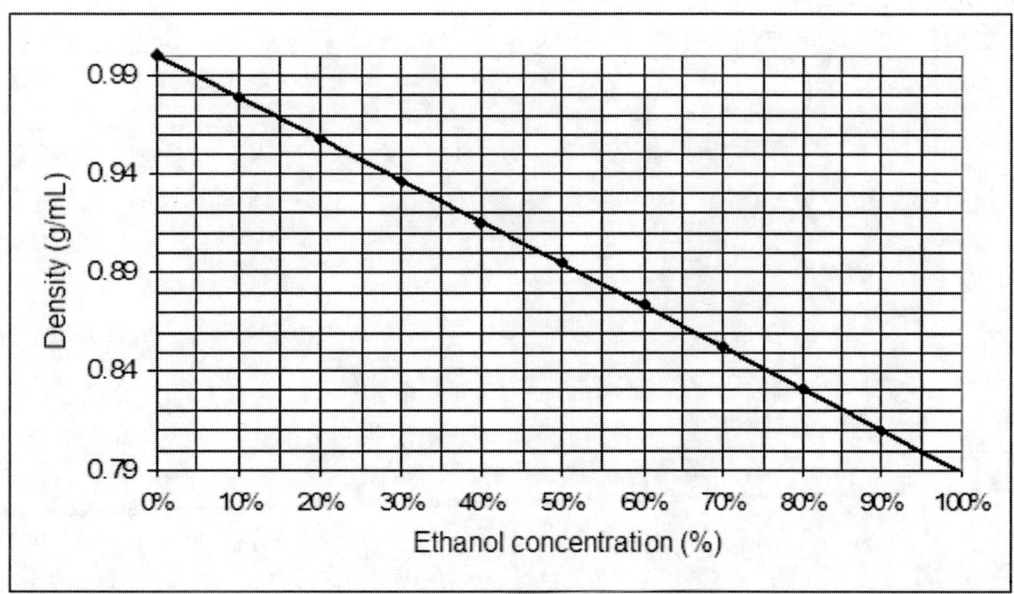

Figure 2. **Density of water-ethanol mixture depending on ethanol concentration**

name   section   date

Pre-Laboratory Assignment

1. Define the following terms, as they relate to this experiment:

  a. Distillation –

  b. Distillate –

  c. Boiling point elevation –

  d. Miscible –

2. The two **white solids** below look very much alike, so you cannot tell them apart.

	Soluble in:	
	Water	Methylene chloride
Caffeine	yes	yes
Sucrose (sugar)	yes	no

Use the solubility information given in the table above to devise a method to separate a mixture of sucrose and caffeine from each other and to end with two dry solids. (HINT: Use 3 methods from the Table 1 on page 111). Write the name of each procedure used and how it is being used in this problem.

For example:

Filtering was then used to separate the dissolved sugar from the solid, undissolved caffeine. (this is not a correct statement for this problem!).

3. A mixture contains 99% acetone and 1% water. What can you say about the boiling point of this mixture, if the boiling point of pure water is 100°C and the boiling point of pure acetone is 57°C? *Explain briefly. (HINT: refer to the Background section of this lab)*

4. What is the function of the ethanol in mouthwash?

<table>
<tr><td>name</td><td>section</td><td>date</td></tr>
</table>

Data Sheet

	Trial 1	*Trial 2*
Mass of empty graduated cylinder, g		
Volume of the distillate, mL		
Mass of cylinder with the distillate, g		
Boiling point of the distillate, °C		

Calculations

Show your work for each calculation! Use rules of significant figures; include units with each result.

	Trial 1	*Trial 2*
1. Calculate the mass of the distillate:	$m =$ ___________	$m =$ ___________
2. Calculate the density of the distillate:	$d =$ ___________	$d =$ ___________

3. Determine the percent ethanol in your distillate by using the chart (in Figure 2):

$\% =$ ___________ $\qquad$ $\% \ =$ ___________

name section date

Post-Laboratory Assignment

1. Based on your boiling point measurements and calculated percent ethanol for the distillate:

    a. Based on the percentage, what were the solute and the solvent in your mixture? *Explain!*

    b. How did the observed boiling point for each trial compare to the boiling points of ethanol (78.3°C) and water (100.0°C)?

    c. Based on the boiling point elevation affect of a solute on a solvent (page 113) what would you have expected to happen to the boiling point of your mixtures? Did the observations match what you would have expected? *Explain!*

2. A student performing one trial of this experiment recovered 4.7 mL of distillate from a mouthwash distillation. The recovered sample had a mass of 3.90 g. What is the percentage of ethanol in the distillate? *(HINT: Refer to Figure 2)*

EXPERIMENT 12: **Synthesis of Esters**

Mike Goldin (Liberty University)

In pouring this fragrant oil on My body, she did it for My burial.
– Matthew 26:12 (NKJV)

Background

Ah, the familiar fragrance of apples, oranges, raspberries, pears, bananas, apricots, spearmint... but did you know that each one of these fruity smells is an ester? As you may recall from Chapter 7 in your lecture textbook, **esters** are carboxylic acid derivatives containing a carboxyl group with a carbon chain attached to the –O of the functional group; as it turns out, you have already smelled many of them! Table 1 below is a partial reproduction of a table from Chapter 10 if your lecture textbook, listing a few ester structures as well as the fragrance that they have.

Table 1. **Definitions of acids and bases**

Ester	Fragrance	Ester	Fragrance
	Banana		Pineapple
	Orange		Grape
	Apple		Wintergreen

In this experiment, you will perform some **esterification** reactions, described in Chapter 10 of the lecture textbook; this is a reaction by which an ester is produced from an alcohol and a carboxylic acid in the presence of a catalyst (typically sulfuric acid or phosphoric acid is used to catalyze this reaction). H_2O is formed as one of the products from the combination of H from the hydroxyl group of the alcohol and the OH from the carboxyl group of the acid, and the remaining organic parts connect to each other. The functional group reaction is shown in a general way in Eq. 1 below:

$$\text{—C(=O)—OH} + \text{HO—C—} \longrightarrow \text{—C(=O)—O—C—} + H_2O \qquad [1]$$

We can also break up esters and get back the carboxylic acid and alcohol. We have to react the acid derivatives with water to do so, and the reaction is called **hydrolysis**. Depending on whether we use an acidic or basic solution to do this, we will get slightly different-looking products (Eq. 2-3).

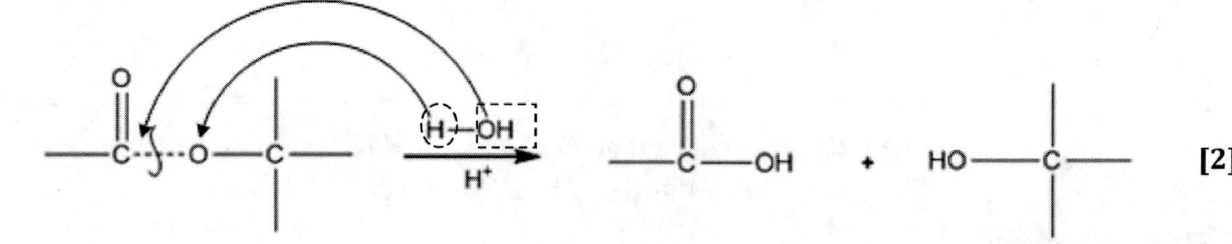

In this experiment you will synthesize five esters from various alcohols and carboxylic acids in acidic conditions (created by adding sulfuric acid, which acts as a catalyst), observing the odors of the reactants and products. Prepare to be surprised by the vast difference in the fragrances before and after the reaction! You will also write structures of the reactants and chemical equations for each of the esterification reactions, as well as name the esters produced.

Objectives

1. To observe the odor of several carboxylic acids, alcohols, and esters
2. To perform esterification reactions of several carboxylic acids and alcohols
3. To write the chemical equations for the esterification reactions in the experiment

Safety

- Wear approved **safety glasses or goggles** during the entire experiment! It is especially important when we are working with toxic, corrosive, and irritant substances.
- Concentrated sulfuric acid, organic acids, and some of the alcohols are toxic, corrosive, and can irritate or cause burns.
- Remember the location of the emergency eyewash and safety shower in case of accidents!
- Wash your hands thoroughly before leaving the laboratory.

Materials and Equipment

You will need the following for your experiment:

1. Acetic acid
2. Butyric acid
3. Formic acid
4. Salicylic acid
5. Ethyl alcohol
6. Isoamyl alcohol
7. Isobutyl alcohol
8. Methyl alcohol
9. Octyl alcohol
10. Sulfuric acid
11. Beaker, 400 mL
12. Hot plate
13. Thermometer, alcohol-based
14. Test tubes (5)
15. Test tube tongs
16. Test tube rack
17. Sharpie or marking tape

Procedure

1. Set up a hot water bath by half-filling a 400-mL beaker with tap water and heating it on the hot plate with a HEAT knob setting of 250°C. Place a thermometer in it and monitor the temperature, adjusting it so that it **does not exceed 70°C** throughout the experiment.

2. With a sharpie or marking tape, mark five test tubes as number 1 through 5.

3. <u>**Under a fume hood**</u>, use the plastic pipette on each carboxylic acid bottle to place <u>**ten drops**</u> of the acid in test tubes as described below:

    *   Ten drops of **acetic acid** (ethanoic acid) into test tube number **1**;
    *   Ten drops of **acetic acid** (ethanoic acid) into test tube number **2**;
    *   Ten drops of **butyric acid** (butanoic acid) into test tube number **3**;
    *   Ten drops of **formic acid** (methanoic acid) into test tube number **4**.

> *<u>**IMPORTANT**</u>: Please use the pipettes (eye droppers) provided with the acid bottles for this purpose! Do not use the same pipette for different acids—this will lead to <u>**contamination**</u> and errors in everyone's observations!*

    Note the odor of each one by wafting the air above the test tube toward your face. Be careful not to inhale too much of the vapor from each one, since all of these acids are rather pungent, and one of them is particularly unpleasant-smelling! Record your odor observations for each on the Data Sheet.

4. From the bottle with salicylic acid, dispense a *small amount* of **salicylic acid** (just enough to cover the end of a spatula or scoopula) into test tube number **5**. Note the odor of salicylic acid by wafting the air above the bottle toward your face. Record your odor observation on the Data Sheet.

5. <u>**Under a fume hood**</u>, use the plastic pipette on each alcohol bottle to dispense precisely **1.00 mL** of the alcohol in a 10-mL graduated cylinder. Note the odor of each alcohol listed below, <u>**EXCEPT FOR METHYL ALCOHOL (which is toxic and SHOULD NOT be inhaled at all!),**</u> by wafting the air above the graduated cylinder toward your face. Be careful not to inhale too much of the vapor from each one, since some of the alcohols do not have a pleasant odor! Record your odor observations for each on the Data Sheet.

    Add the 1.00 mL of alcohol to the appropriate test tube (that already contains a carboxylic acid from Step 3) as described below:

    *   1.00 mL of **isoamyl alcohol** (3-methyl-1-butanol) into test tube number **1**;
    *   1.00 mL of **octyl alcohol** (1-octanol) into test tube number **2**;
    *   1.00 mL of **ethyl alcohol** (ethanol) into test tube number **3**;
    *   1.00 mL of **isobutyl alcohol** (2-methyl-1-propanol) into test tube number **4**;
    *   1.00 mL of **methyl alcohol** (methanol) into test tube number **5**.

> *<u>**IMPORTANT**</u>: Please use the pipettes (eye droppers) provided with the alcohol bottles for this purpose! Do not use the same pipette for different alcohols—this will lead to <u>**contamination**</u> and errors in everyone's observations!*

6. Add **five (5) drops** of concentrated sulfuric acid to each of the test tubes.

> ***DANGER!*** *Concentrated sulfuric acid is toxic and corrosive! It will burn your skin, eyes, or clothes if you get it on you! As always, it is **imperative** that you keep your safety glasses on **at all times** while anyone is still conducting their experiment – even if you are finished with yours. Wearing gloves is highly recommended when working with H_2SO_4.*

7. Place the test tubes into the 70°C water bath for 15 min to allow the reaction to occur and the products to form.

8. Using test tube tongs, carefully take the test tubes out of the hot water bath and place them in the test tube rack. One by one, lift them out of the rack with test tube tongs and waft the air above the test tube toward your face to observe the odor of the product. Record your odor observations for each on the Data Sheet.

Clean-Up

- Clean up any spills with soap and water. Do this as soon as you find a spill!

- Discard the contents of each test tube into the collection container labeled "MIXED ESTERS, CARBOXYLIC ACIDS, AND ALCOHOLS," taking care not to get them on your hands.

- Wash all of the test tubes with soap and water using the test tube brushes located by each of the two large sinks.

- Return the thermometer to the front desk where you borrowed it from!

name	section	date

Pre-Laboratory Assignment

1. Define the terms below. *(You may need to refer back to previous experiments or your textbook.)*

    a. Carboxyl group –

    b. Hydroxyl group –

    c. Ester –

    d. Esterification –

    e. Hydrolysis –

2. Draw condensed structural formulas or skeletal line structures of the carboxylic acids and alcohols below. Look the structure up online if you cannot find it elsewhere.

Acid Name	*Structure*	*Alcohol Name*	*Structure*
Ethanoic acid (acetic acid)		Ethanol (ethyl alcohol)	
Butanoic acid (butyric acid)		Methanol (methyl alcohol)	
Methanoic acid (formic acid)		1-Octanol (octyl alcohol)	
Salicylic acid		3-Methyl-1-butanol (isoamyl alcohol)	
Sulfuric acid		2-Methyl-1-propanol (isobutyl alcohol)	

3. Write the products for the following hydrolysis reactions. *HINT:* Follow examples on page 147 and 148 of the Background section.

a) $CH_3-\underset{\underset{\displaystyle }{\|}}{\overset{\overset{\displaystyle O}{\|}}{C}}-O-CH_2-CH_3 \quad \xrightarrow[\text{H}^+]{\text{H}_2\text{O}}$

b) $CH_3-\underset{\underset{\displaystyle }{\|}}{\overset{\overset{\displaystyle O}{\|}}{C}}-O-CH_2-CH_3 \quad \xrightarrow{\text{NaOH}}$

name ____________________ section __________ date __________

Data/Observation Sheet

I. Odors of the reactants:

Acid Name	*Odor*	*Alcohol Name*	*Odor*
Acetic acid		Isoamyl alcohol	
		Octyl alcohol	
Butyric acid		Ethyl alcohol	
Formic acid		Isobutyl alcohol	
Salicylic acid		Methyl alcohol	

II. Esterification products:

Test Tube #	*Odor*	*Ester Structure**	*Ester Name**
1			
2			
3			
4			
5			

* You will complete these two columns as a part of your Post-Laboratory Assignment

Chemical Equation Worksheet

Write the chemical equations for the esterification reactions that occurred in Test Tubes 1–5. Use the structural formulas from the Pre-Lab and the Procedure directions on which acids and alcohols are placed together in each test tube to help you figure out the reactants. Use Equation 1 to help with the pattern of how the products are formed.

Test Tube 1:

Test Tube 2:

Test Tube 3:

Test Tube 4:

Test Tube 5:

____   ________   ________

name   section   date

Post-Laboratory Assignment

1. Using your Chemical Equation Worksheet, fill in the ester structures in the second table on the Data/Observation Sheet page. Then, provide the name of the ester. *(HINT: You may want to use the common names of the alcohols to help you name the esters. It would be a bit more complicated if you do not!)*

2. Using what you have learned about esterification reactions in this experiment, complete the chemical equation for the following esterification reaction:

$$CH_3-CH_2-OH \;+\; HO-\overset{\overset{\textstyle O}{\|}}{C}-CH_3 \;\underset{\longleftarrow}{\overset{H^+,\,heat}{\longrightarrow}}$$

3. **Amidation** is a reaction similar to esterification, except an amine is added to the carboxylic acid instead of an alcohol, producing an amide. Using the pattern you have learned from esterification (and information in your textbook), complete the chemical equation for the following amidation reaction:

$$CH_3-CH_2-NH_2 \;+\; HO-\overset{\overset{\textstyle O}{\|}}{C}-CH_2-CH_3 \;\underset{\longleftarrow}{\overset{heat}{\longrightarrow}}$$

EXPERIMENT 13: **Separation of Milk into Its Components**

Mark Hemric, Mike Goldin (Liberty University)

As newborn babes, desire the pure milk of the word, that you may grow thereby.
– 1 Peter 2:2 (NKJV)

Background

The American dairy farmers owe a debt of gratitude to three men: Dr. John H. Kellogg, his brother W.K. Kellogg, and C.W. Post. Between the three of them, America's breakfast food of choice—cold cereal—was invented and popularized. Cereal and milk in the morning are as much a part of American life as Stars and Stripes, Mom, and apple pie. One might even think that God had cereal in mind when he created milk…

But what is the real purpose of milk? Of course, it was created by God as part of His provision for newborn babies—human ones, as well as the young of mammals. It is very obvious just how much God cares for these young lives if we consider how important milk is to their health and well-being. Here are just a few aspects:

- Milk contains nutrition in the form of carbohydrates, proteins, and fat in the amounts that are "just right" for the growing baby;

- It also contains digestive proteins to help "jump-start" the baby's digestive system, as well as minerals, vitamins, and hormones that the baby needs;

- The antibodies from the mother that are also present in milk help the baby resist infections while his own immune system is too weak;

- Each time the new mother is breastfeeding, her milk changes slightly in content: at the beginning it contains more proteins and lactose (milk sugar) and is more watery, to satisfy the baby's thirst; at the end of the feeding session, it contains more fat and is therefore "creamier," helping the baby feel fuller and helping him gain weight;

- As the baby grows, the content of the milk will adjust to some extent with his changing needs.

As a baby grows, he usually loses ability to digest milk partly or completely. Therefore, we can say that most people are at least somewhat lactose-intolerant—although for some, this intolerance may be quite severe!

As mentioned above, milk contains carbohydrates (sugars), proteins, and lipids (fats), as well as calcium and phosphate ions. These are three of the four main types of biochemical compounds. In lecture, we will look in much greater detail at these types of compounds. Here is a typical **carbohydrate**, glucose (blood sugar):

Both drawings are a representation of the same molecule, except in the one on the right the glucose molecule is closed in a cyclic ring structure. Notice that there are several OH groups, which makes it polar and soluble in water. In fact, all sugars are polar, including **lactose** (milk sugar):

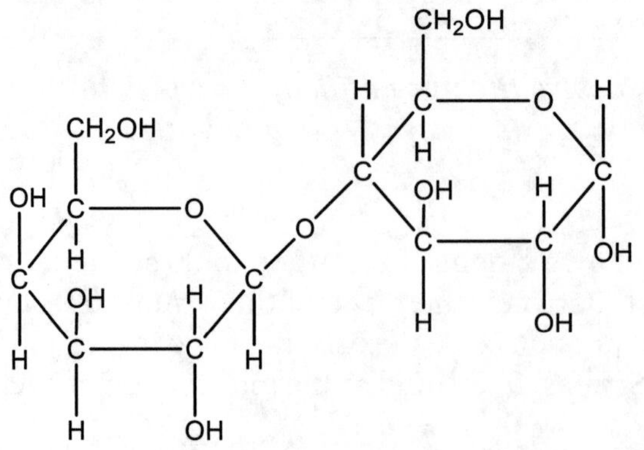

The other two types of biochemical compounds are both longer carbon chains. **Lipids** contain carboxyl groups attached to long carbon chains and are generally nonpolar because of how long the carbon chains are. Here is a typical triglyceride, a type of lipid:

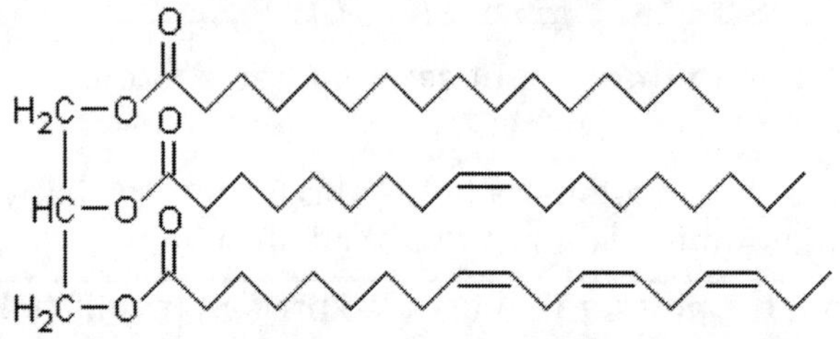

Proteins generally also contain lots of carbons connected together. However, they have many different functional groups attached to the main carbon chain, plus there are polar carboxyl and amino groups connected to each other. Therefore, smaller proteins are quite polar and will dissolve in water. Large proteins are not as soluble in water due to their size. Their solubility in water is greatly affected by the interactions (attractions!) between different functional groups within the same molecule. Usually these attractions hold many of the protein molecules in a globular shape. If they are disrupted by a change in acidity, heating, vigorous stirring, etc., the large proteins can change their shape and become more "stringy" and less soluble. We call this disruption in protein structure **denaturation**. In fact, the reason milk curdles when it is unrefrigerated (or stays in your fridge too long!) is because the small amount of bacteria in milk that remains after pasteurization produce acid, causing the proteins to denature.

When we separate milk into its components in this experiment, we denature its proteins by adding acid and then heating it. When the insoluble denatured large proteins come out of solution, they take the nonpolar lipids out of solution with them. We then further separate the lipids from the large proteins and perform chemical tests to confirm the presence of each type of biochemical compound. To determine whether a solution contains calcium and phosphate ions, we use two precipitation reactions – with aqueous ammonium oxalate, $(NH_4)_2C_2O_4$, for calcium and with aqueous ammonium molybdate, $(NH_4)_2MoO_4$, dissolved in nitric acid for phosphate. Their net ionic equations are shown in Eq. 1-2:

$$Ca^{2+}(aq) + C_2O_4^{2-}(aq) \rightarrow \textbf{CaC}_2\textbf{O}_4\textbf{(s, white)} \tag{1}$$

$$PO_4^{3-}(aq) + 12MoO_4^{2-}(aq) + 3NH_4^+(aq) + 24H^+(aq) \xrightarrow{HNO_3} \textbf{(NH}_4\textbf{)}_3\textbf{PO}_4\cdot\textbf{12MoO}_3\textbf{(s, yellow)} + 12H_2O(l) \tag{2}$$

While Eq. 2 looks (and is!) quite complicated, the bottom line is that the yellow precipitate, called ammonium phosphomolybdate, appears when PO_4^{3-} ions are present in solution.

For determining whether a solution contains lactose, we can use the so-called Benedict's test, in which a solution containing copper(II) ions reacts with an aldehyde group of the sugar molecule, oxidizing it to a carboxylic acid, while the copper(II) ion is reduced to copper(I) in the form of an insoluble brick-red copper(I) oxide precipitate, Cu_2O. This, of course, can only happen if the lactose has opened up into its "open-chain" form rather than its Haworth-structure "closed-ring" furanose form. The reaction of lactose oxidation by Benedict's reagent is shown in Eq. 3:

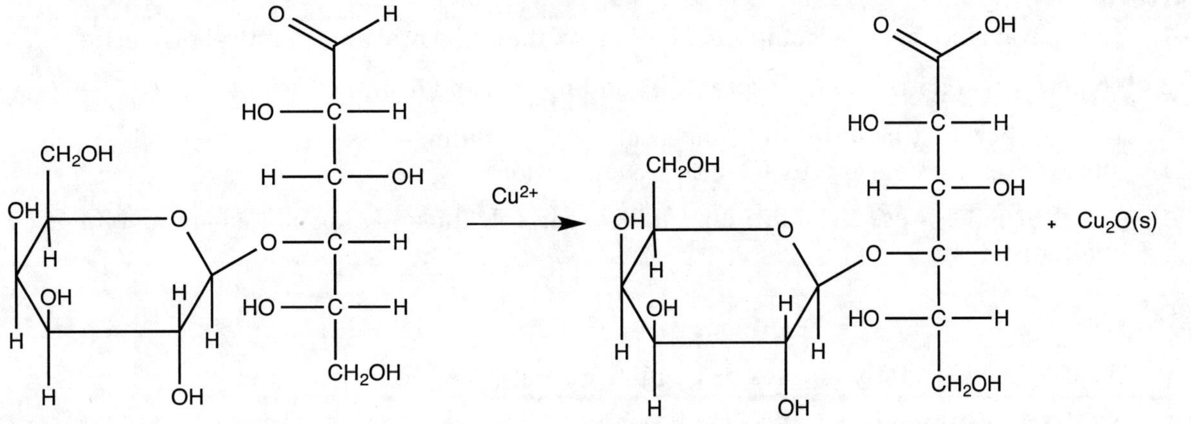

$$[3]$$

By the way, since Benedict's test works by oxidizing the aldehyde group, this means that Benedict's reagent cannot be used on ketoses, since they have no aldehyde group. In fact, this is why aldoses are also called "reducing sugars," since they can be oxidized and therefore act as a reducing agent.

For detecting the presence of lipids (milkfat), which will first be separated by extracting them from the mixture with methylene chloride, a simple test for the presence of unsaturated carbon chains is performed by using a bromimation addition reaction. The C=C double bonds in the unsaturated fatty acids that are part of the triglycerides in milk will react with a molecule of bromine, Br_2, dissolved in methylene chloride, according to Eq. 4. Only the double bond is shown in this equation, rather than the entire triglyceride molecule.)

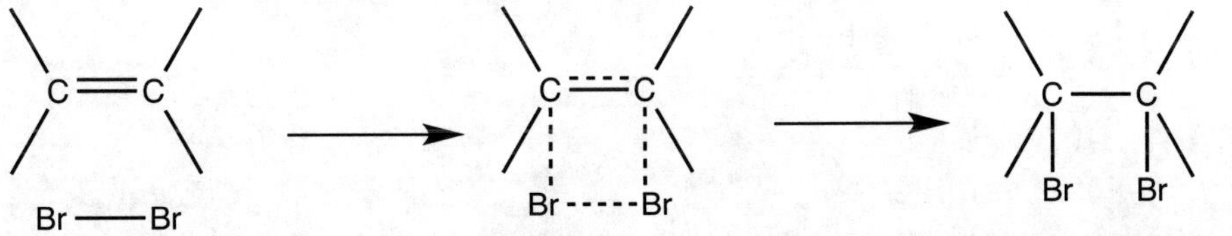

$$[4]$$

As this reaction occurs, the orange-brown color of the dilute bromine solution will change to clear, as Br_2 is used up and added to the lipid molecules. If the brown/orange color remains, it would actually not imply the absence of lipids from solution, but only that there are no unsaturated carbon chains; however, since milkfat always contains some amount of unsaturated fatty acids, this is a pretty good indication of whether milkfat is in the methylene chloride after the extraction.

Finally, the protein test is a commercially produced Bradford Protein Assay, which contains the molecule shown in Figure 1 that changes color from a brownish-red to a bright blue color when it reacts with a protein molecule. As you can see, it is quite complex, and it is not necessary for our purposes to indicate exactly what functional groups in the molecule are responsible for

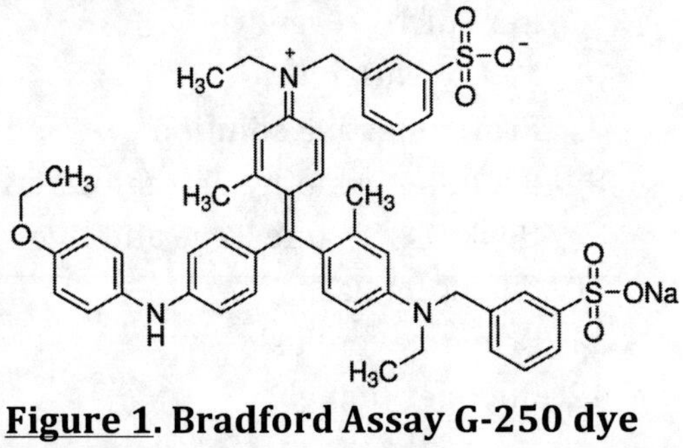

Figure 1. Bradford Assay G-250 dye

the color change. However, what is important to know is that the color of this molecule also depends on pH, and slightly acidic water (such as tap water) can change the color to a pale greenish-blue or aqua color. Therefore, it is important not to add excessive amounts of the sample solution (or water, which is used for comparison of color) and to follow the direction in Procedure closely.

Objectives

1. To separate biochemical compounds based on their physical and chemical properties

2. To perform denaturation of proteins by adding acid and heating

3. To use laboratory skills in, and conceptual understanding of, separation methods: vacuum filtration, extraction, gravity filtration, evaporation

4. To confirm the presence of proteins, milk sugar, calcium and phosphate ions, and lipids by qualitative tests

Safety

- Wear approved **safety glasses or goggles** during the entire experiment!

- Avoid **burns** by contact with hot objects! Do not hand hot objects to another person!

- Hydrochloric acid can cause burns and irritation of your skin and is **dangerous** to your eyes! Wash it off if any of it gets on your skin, and **use the eyewash if you get it in your eyes!**

- Methylene chloride is **toxic**! Do not get it on skin; wash off quickly if you do! **Wear gloves!** Only use it **UNDER THE HOOD** and **NEVER INHALE ITS FUMES!**

- Bromine dissolved in methylene chloride is corrosive and an irritant. This is in addition to the hazards from methylene chloride in which it is dissolved! **Wear gloves!** Only use it **UNDER THE HOOD** and **NEVER INHALE ITS FUMES!**

- Be careful with the testing solutions! Wash off your skin if you get any on you.

- Wash hands thoroughly before leaving the laboratory!

Materials and Equipment

You will need the following for your experiment:

1. Milk	13. Rubber funnel adaptor *
2. 3 M hydrochloric acid	14. Rubber hose *
3. 0.1 M ammonium oxalate	15. Filter paper (2 pieces) *
4. 0.5 M ammonium molybdate	16. Wash bottle with water
5. Benedict's reagent	17. Glass rod with rubber policeman
6. Methylene chloride	18. Hot plate
7. Protein testing solution	19. Crucible tongs (or hot mitt) *
8. Bromine in methylene chloride sol'n	20. Boiling stones
9. Beakers (3 – one large and two small)	21. Thermometer, alcohol-based *
10. Filtration flask	22. Test tube (large) *
11. Ernenmeyer flask	23. Plastic pipette
12. Buchner funnel *	

Procedure

Below is a diagram of the steps in the experiment (Figure 2).

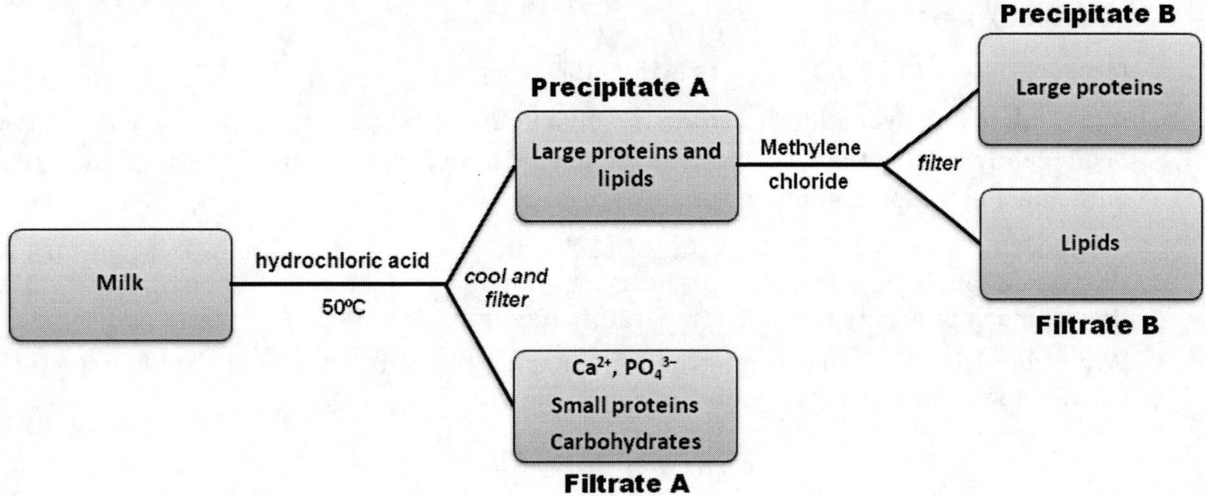

Figure 2. A diagram of the experimental steps to separate the components of milk

Table 1 details the two major separation steps you will do, while Table 2 summarizes which qualitative chemical tests are used for which parts:

Table 1. Reagents for milk separation

Reagent	Part	Added to	Filtrate produced	Precipitate residue
Hydrochloric acid (+ heat)	A	Original milk	**Filtrate A:** Ca^{2+}, PO_4^{3-}, small proteins, carbohydrates	**Precipitate A:** Large proteins, lipids
Methylene chloride	D	Precipitate A	**Filtrate B:** Lipids	**Precipitate B:** Large proteins

Table 2. Chemical identification tests for various components of milk.

Reagent	Part	To Determine	Precipitate A	Filtrate A	Precipitate B	Filtrate B	Water
Ammonium oxalate	B	Ca^{2+}		✓			
Ammonium molybdate	B	PO_4^{3-}		✓			
Benedict's reagent	C	Reducing sugars		✓			
Protein Testing Sol'n	D	Proteins		✓	✓		✓
Cooling (ice bath)	E	Saturated/ unsat. fats				✓	

Below are the detailed directions for each step:

1. Obtain the equipment marked in the list above with an asterisk "*" from the front desk.

A. **Making Milk Whey:**

2. Measure out about 25 mL milk into a 150 mL beaker.

3. Add 3 mL of 3 M hydrochloric acid to the milk in the beaker. Let it sit for about 1 min, then begin gently leaning the beaker from side to side in order to mix it. Allow the acid to react with milk for 1-2 more minutes.

4. Set your hot plate to a "HEAT" setting of 120°C and place the beaker on it. Heat the milk for **_no more_** than 10 minutes. As you heat, monitor the temperature in your beaker with a thermometer to ensure the milk temperature **never** exceeds 50°C.

5. While one lab partner is heating and monitoring the milk, the other should set up the vacuum filtration apparatus, shown in Fig. 3 below:

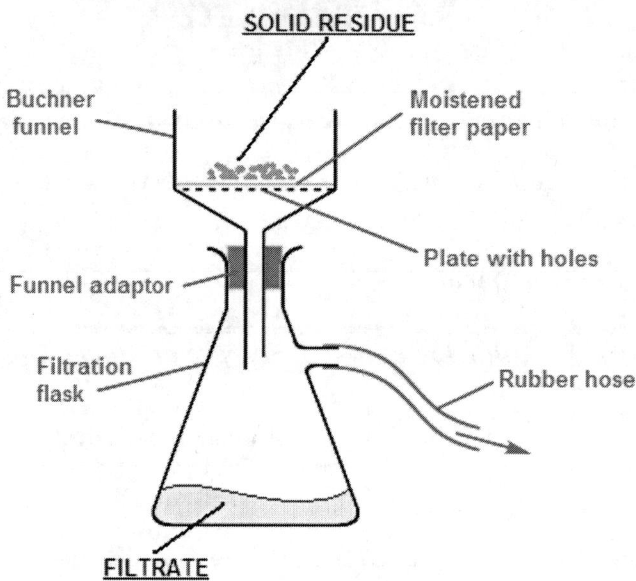

Figure 3. Vacuum filtration assembly.

- Carefully, but firmly, clamp the filtration flask to the metal rod at your work bench;
- Attach one end of the rubber hose to the "side arm" of the flask;
- Attach the other end of the rubber hose to the "VAC" valve on your bench top;
- Place a rubber funnel adaptor in the mouth of the flask;
- Place the Buchner funnel on top of the adaptor and flask (as shown in Fig. 3);
- Place the filter paper inside the funnel;
- Turn on the vacuum (by turning the valve handle parallel with the nozzle);

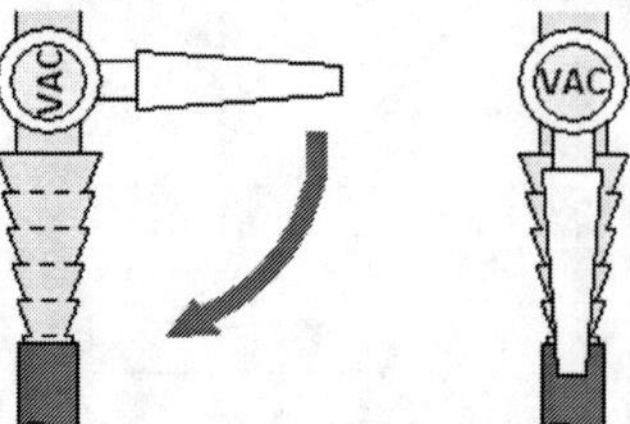

- Wet the filter paper completely by squirting water from the wash bottle, so there is an airtight seal between the funnel and paper. You can turn off the vacuum until it is needed.

6. After 10 minutes, remove the beaker from the hot plate to cool (running cold water over the bottom of the beaker speeds this up!)

7. Place a 250- or 400-mL beaker, half full of warm water from the faucet, onto the hot plate in preparation for Sections C and D and **increase your hot plate to a setting of 300°C**.

8. Turn on the vacuum in your vacuum filtration assembly. Pour the cooled milk solution into the Buchner funnel to separate the protein/fat solid (Precipitate A) from the ion/carbohydrate liquid (Filtrate A) by filtration. After 10-15 minutes, several milliliters of liquid (Filtrate A) should collect.

9. Turn off the vacuum, disconnect the hose to the side of the flask, and remove the Buchner funnel from the flask. Pour the liquid (Filtrate A) into a beaker and **save for testing (Sections B-D)**. Replace the Buchner funnel and resume vacuum filtering.

10. Now, test your liquid (Filtrate A) for ions and sugar (**Section B and C**). (**Note: save the remaining filtrate A for the protein test – Section D**)

B. <u>Testing for the Ions, Calcium and Phosphate:</u>

11. <u>Calcium:</u>   Add 1 mL of 0.1 M ammonium oxalate, $(NH_4)_2C_2O_4$, and 1 mL of Filtrate A to a small test tube. If a white precipitate (CaC_2O_4, calcium oxalate) forms, this indicates the presence of calcium ions in milk.

12. <u>Phosphate:</u>   Add 1 mL of 0.5 M ammonium molybdate, $(NH_4)_2MoO_4$, and 1 mL of Filtrate A to a small test tube. If a yellow precipitate ($(NH_4)_3PO_4 \cdot 12\ MoO_3$, ammonium phospho-molybdate) forms, this indicates the presence of phosphate ions in milk.

C. <u>Testing for the Sugar:</u> Use Benedict's test to prove the presence of a reducing sugar, the disaccharide lactose.

13. Add 1 mL of Benedict's reagent to a small test tube.

14. Add 4 to 8 drops of Filtrate A and place the test tube in your hot water bath. Monitor the color inside the test tube for 10 minutes or more; the color will change gradually! If you have a sugar molecule, you will see the copper reduced from a "blue" ion to a copper-red orange precipitate. (*This was once used to test for blood sugar, glucose, in the urine of diabetics.*)

D. <u>Testing for the Protein:</u>

15. Turn off the vacuum, disconnect the hose from your Buchner funnel, and remove the filter paper from the funnel with your spatula.

16. Gently scrape your protein/lipid solid (Precipitate A) off the filter paper and into your 50- or 100-mL beaker. Throw away the used filter paper in the trash can.

17. **UNDER THE HOOD,** add 10 mL of methylene chloride (CH_2Cl_2, dichloromethane) to the 50-mL beaker and mix thoroughly with your stirring rod.

> **<u>IMPORTANT</u>: Be sure to reclose the bottle of the TOXIC methylene chloride!! Keep any methylene chloride solutions under the hood!**

18. Wipe out your white Buchner funnel with a **<u>dry</u>** paper towel. ***DO NOT use any water!*** Place it on a ***<u>clean, dry</u>*** 250-mL (or 125-mL) Erlenmeyer flask.

19. Put a piece of filter paper in the Buchner funnel. Since you are using the Erlenmeyer flask, you will not need the vacuum for this part!

20. **UNDER THE HOOD, SLOWLY** pour the solution from Step 17 into the Buchner funnel to separate the protein solid (Precipitate B) from the liquid containing your lipid (Filtrate B). **Save the liquid (Filtrate B) for the lipid test – Section E.**

21. Scrape a small (pea-sized) amount of Precipitate B into a test tube and add about 10 mL of **0.1 M NaCl**. Stir with a glass rod or spatula until the solution is *cloudy*! (The protein should at least partially dissolve into the salt solution).

22. You will be testing for protein in these three samples: Large Proteins (as prepared in Step 21 above), Small Proteins (Filtrate A) and water (a "blank" solution with no protein in it!).

    a.   Label 3 **clean** test tubes as "W", "LP" and "SP."

    b.   Put 5 drops of water into the "W" test tube.

    c.   Put 5 drops of the large protein solution (prepared in Step 21) in the "LP" test tube.

    d.   Put 5 drops of the small protein solution (Filtrate A) in the "SP" test tube.

    NOTE: make sure you rinse your pipette/eye dropper between samples!

23. Add 2 drops of protein-testing solution into each of the 3 test tubes.

    Is there protein in your sample? If so, the test tubes containing 5 drops of LP or SP will turn **dark blue** compared to the test tube labeled "W."

E. Testing for Fat: *USE GLOVES; KEEP UNDER HOOD DUE TO <u>TOXIC</u> METHYLENE CHLORIDE!*

24. **UNDER THE HOOD,** pour about 1 mL of the liquid Filtrate B (from Step 20) from the Erlenmeyer flask into a test tube.

25. **UNDER THE HOOD,** use the pipette designated for the bromine solution to add a few drops of the brown/orange bromine solution into the test tube with Filtrate B.

> *<u>WARNING</u>: Bromine is corrosive and will irritate skin, eyes, or digestive tract. It is dissolved in toxic methylene chloride. DO NOT allow any contact with skin; wear gloves when performing this part of the experiment!*

If the brown/orange color disappears and the Filtrate B solution returns to colorless, this indicates the presence of unsaturated fat.

Clean-Up

- **PUT ANY REMAINING METHYLENE CHLORIDE AND THE REMAINING *FILTRATE B* IN THE "DISCARD METHYLENE CHLORIDE" CONTAINER <u>UNDER THE HOOD</u>!**

- **PLACE THE FILTER PAPER AND REMAINING *PRECIPITATE B* (PART B) IN THE "DISCARD SOLIDS" CONTAINER UNDER THE HOOD! (It contains toxic fumes.)**

- Place the contents of the test tubes with chemical tests from Parts B, C, D into designated disposal containers at the front desk

- Wash any greasy or dirty glassware (flasks, test tubes, etc.) with plenty of soap in the sink

- Wash off any stains from the lab benches with a soapy paper towel. Dry the bench tops with a dry paper towel.

- Return the Buchner funnel and rubber funnel adaptor, rubber hose, crucible tongs or hot mitt, thermometer, and large test tube to the front desk

- Wash your hands thoroughly with soap before leaving the lab!

---

name   section   date

Pre-Laboratory Assignment

1. Which of these types of biochemical compounds are polar, and which are nonpolar? Which are soluble in water? Complete the table below.

    Use "P" for polar or "N" for nonpolar. Use "S" for soluble or "I" for insoluble in water.

Family	*Polar?*	*Soluble in H_2O?*
Lipids (fats)		
Carbohydrates (sugars)		
Small proteins		
Large proteins		

2. Methylene chloride ("proper" IUPAC name = dichloromethane) is considered a nonpolar solvent. Which of the four types of compounds listed above should dissolve in methylene chloride? Explain.

3. What family of biochemical compounds discussed in this experiment undergoes denaturation? What happens to these compounds when they are denatured?

name   section   date

name   section   date

Data Sheet

A. Making Milk Whey

(1) What did you observe happening to milk after you added hydrochloric acid?

(2) When milk "goes sour" in your refrigerator, what organism is adding acid to your milk?

(3) Which part of milk (proteins, sugars, fats, calcium ions, or phosphate ions) "curdles"? What is the molecular disruption called that causes curdling?

B-D. Tests for Ca^{2+}, PO_4^{3-}, Sugar, Proteins

Describe what you observed when you added the following solutions to **Filtrate A**:

		Change Observed	*Indicates Presence of...*
(4)	0.1 M ammonium oxalate		
(5)	0.5 M ammonium molybdate		
(6)	Benedict's solution		

Describe what you observed when you added the **Protein Testing Solution** to the following solutions:

		Change Observed	*Indicates Presence of...*
(7)	Precip. B + 0.1 M NaCl (LP)		
(8)	Filtrate A (SP)		
(9)	Pure (tap) water (W)		

E. Test for Fats

		Change Observed	*Indicates Presence of...*
(10)	Bromine solution in CH_2Cl_2		

name section date

Post-Laboratory Assignment

1. What parts of the human body are made of calcium phosphate ($Ca_3(PO_4)_2$), which is also found in milk?

2. Sugars are either aldehydes or ketones with a number of OH groups on the parent chain. The so-called "reducing sugars" or *aldoses*, such as lactose in this experiment, contain the aldehyde (–CHO) group and react with Benedict's reagent (similar to regular aldehydes).

   Complete the chemical equation for the reaction of another reducing sugar, glucose, by writing the structure of the product of this reaction. (*HINT*: Look up the reaction of aldehydes with Benedict's reagent to help you, or Equation 3 on page 155 of this lab book.)

$$
\begin{array}{l}
O{=}\overset{1}{C}{-}H \\
H{-}\overset{2}{C}{-}OH \\
HO{-}\overset{3}{C}{-}H \\
H{-}\overset{4}{C}{-}OH \\
H{-}\overset{5}{C}{-}OH \\
\overset{6}{C}H_2OH
\end{array}
\quad + \quad Cu^{2+} \quad \rightarrow \quad \boxed{} \quad + \quad Cu^{+}
$$

3. The different components of milk are separated in this experiment. What physical and/or chemical properties of the components are used to separate them? *List at least two and explain!*

EXPERIMENT 14: **Inventory and Clean-Up**

Mike Goldin (Liberty University)

To everything there is a season, a time for every purpose under heaven.
– Ecclesiastes 3:1 (NKJV)

Congratulations! You have made it through the Chemistry 107 lab sequence! It has been a pleasure going through the labs with you this semester—and I hope you also enjoyed it, at least somewhat!!

Today's meeting is simply to clean up any messes you might still have in your area and to make an inventory of the equipment at your stations. Thank you for doing this work, as it helps both your professor and the Biology/Chemistry Department as a whole. This also helps the students that will use this laboratory facility after you!

This is also an excellent opportunity for you to follow up on any grading concerns or ask questions about the course material in preparation for the final exam.

Directions

Please accomplish the following tasks or any others indicated by the instructors:

1. If there are any spills or stains on the bench top or in your immediate area, please clean them up with a paper towel and a little soap. Then sweep off the bench top with a wet paper towel and use a dry paper towel to dry it off.

2. If your immediate area is clean, please find a common area that might need the same kind of attention.

3. Restock each drawer and cabinet in your station with the items that are indicated on the labels. If a drawer contains an item that is not supposed to be there, put it in the correct drawer or at a common stock area on the side bench indicated by a professor. If something is missing from the drawer, obtain that item from the side bench or cabinet stock.

4. While stocking the stations, please inspect the equipment and glassware for any stains or cracks. Clean any dirty glassware or equipment. Also, run a wet paper towel down the bottom of your drawer. If necessary, clean up any spills inside the drawer with a little soap and a wet paper towel. Dry the bottom of the drawer with a dry paper towel and replace all of the equipment inside.

Mike Goldin (Liberty University)

Does a spring send forth fresh water and <u>bitter</u> from the same opening?
– James 3:11 (NKJV)

Background

What do lemons, buttermilk, and Shock tarts have in common? Right—the sour taste! Sour foods, vinegar, the sting of ants, and gastric juice are all **acidic**—they contain a substance that produces hydrogen ions (H^+, also called protons) when dissolved in water. On the other hand, foods that are **basic** (or **alkaline**) tend to taste bitter; some alkaline foods include broccoli, cabbage, garlic, green beans, and many other vegetables. Bases produce hydroxide ions (OH^-) when they dissolve in water; their solutions usually feel slippery. **(While we talk about the taste and feel of acids and bases, you should NEVER taste acids and bases or allow unprotected skin to come in contact with it! Both strong acids and strong bases can cause severe burns that permanently damage your skin, eyes, or any other part of your body they come in contact with.)**

There are several ways acids and bases are described in chemistry. The two most important ones are the Arrhenius definition and the Brønsted-Lowry definition. These definitions are summarized in Table 1 below.

Table 1. **Definitions of acids and bases**

	Arrhenius Definition	*Brønsted-Lowry Definition*
Acid	Produces H^+ ions when dissolved in water	Donates (gives away) protons, H^+
Base	Produces OH^- ions when dissolved in water	Accepts (takes on) protons, H^+

The **Arrhenius definition** deals more with what ions are produced when acids and bases dissolve in water. For example, a typical Arrhenius acid would contain hydrogen ions and dissociate in water, producing "free-floating" H^+ ions in solution, as shown for hydrochloric acid, HCl, in Eq. 1:

$$HCl(aq) \rightarrow H^+(aq) + Cl^-(aq) \tag{1}$$

Similarly, a typical Arrhenius base would contain hydroxide ions and dissociate in water, producing OH^- ions in solution, as you can see for sodium hydroxide, NaOH, in Eq. 2:

$$NaOH(s) \rightarrow Na^+(aq) + OH^-(aq) \tag{2}$$

If an Arrhenius acid completely reacts with an Arrhenius base in equal amounts, the resulting solution is neither acidic nor basic; we call it **neutral**. The products of such a **neutralization reaction** are water and an ionic compound called a **salt** that is neither acidic nor basic and has the cation that comes from the base and the anion that comes from the acid. You'll notice that neutralization is really a form of double displacement (Eq. 3):

$$NaOH(aq) + HCl(aq) \rightarrow NaCl(aq) + HOH(\ell) \tag{3}$$

Indeed, the H^+ and OH^- end up together as HOH (H_2O), and the remaining Na^+ and Cl^- also end up together.

The **Brønsted-Lowry definition** is a more inclusive one, not tied to what ions are formed in solution, but rather to how compounds react with each other *in a specific reaction*. In fact, acids and bases are related through the transfer of protons (H^+) from acids to bases. For example, in the chemical equation below (Eq. 4):

$$NH_3(aq) + H_2O(\ell) \rightarrow NH_4^+(aq) + OH^-(aq) \qquad [4]$$

$$H^+$$

H_2O donates a proton to ammonia, NH_3. H_2O becomes OH^- by losing the H^+, and NH_3 becomes NH_4^+ by gaining the H^+. In fact, we can describe an acid-base reaction according to the Brønsted-Lowry definition as a **proton (H^+) transfer** reaction.

According to the definitions above, H_2O is an acid here, since it donated H^+; NH_3 is a base, since it accepted the H^+ from water. By the way, notice that NH_3 would also be a base according to the Arrhenius definition—it really doesn't require a compound to *contain* OH^-, only to *produce it when dissolved in water*. Since NH_3 reacts with water, producing OH^-, it fits this definition of a base. In the same way, we can describe HCl as an acid by writing its reaction with water like this (Eq. 5):

$$HCl(aq) + H_2O(\ell) \rightarrow H_3O^+(aq) + Cl^-(aq) \qquad [5]$$

In Eq. 4, you can see that HCl donates its proton to H_2O. HCl becomes Cl^-, and H_2O gains a proton and becomes H_3O^+. This cation, H_3O^+, is called **hydronium**; it is simply a proton attached to a water molecule. In other words, *since H^+ and H_3O^+ are the same thing*, we can also say that Eq. 1 and Eq. 5 are essentially the same, the only difference being that H_2O is omitted from Eq. 1. In fact, a proton H^+ is so small that it *always* attaches to a water molecule in solution and exists in the form of H_3O^+, even though we might write "H^+" instead. We will sometimes use "H^+" and "H_3O^+" interchangeably.

Here is another example of a Brønsted-Lowry acid, the ammonium ion, NH_4^+:

$$NH_4^+(aq) + H_2O(\ell) \rightarrow NH_3(aq) + H_3O^+(aq) \qquad [6]$$

Now that we understand the Brønsted-Lowry definition of acids and bases, we can discuss one important quality of water. Water is neither acidic nor basic, but *it can act either as an acid or a base, depending on its environment.* In fact, water can react with itself, where one water molecule acts as an acid and the other as a base (as seen in Eq. 7 below):

$$H_2O(\ell) + H_2O(\ell) \rightleftharpoons H_3O^+(aq) + OH^-(aq) \qquad [7]$$

$$H^+$$

Since we can write "H^+" instead of "H_3O^+", Eq. 7 can also be written like this:

$$H_2O(\ell) \rightleftharpoons H^+(aq) + OH^-(aq) \qquad [8]$$

Both Eq. 7 and 8 represent a *reversible reaction*, indicating that chemical *equilibrium* is reached after some time. Let's think about what happens if this equilibrium is disturbed:

- **If an acid is added:** more H^+ (a.k.a. H_3O^+) is added to the mixture—and H^+ is a ***product***! Adding a product will shift the equilibrium to the ***left*** (towards reactants), which means the amount of H_2O will increase and the amounts of both H^+ and OH^- will decrease. This means that ***adding H^+ will decrease the amount of OH^- in solution***.

- **If a base is added:** more OH^- is added to the mixture—which is also a ***product***! Adding a product, again, will shift the equilibrium to the ***left*** (towards reactants), which means the amount of H_2O will increase and the amounts of both H^+ and OH^- will decrease. Likewise, this means that ***adding OH^- will decrease the amount of H^+ in solution***.

In a practical sense, then, H^+ and OH^- are "opposites": when one is increased, the other one decreases. Since water always produces both H^+ and OH^- as discussed above, we will always have some of each in an aqueous solution. When we have more H^+ (or we could say the H^+ *concentration is higher*) than OH^-, the solution is acidic; when we have more OH^- than H^+, the solution is basic.

Usually we measure concentrations of aqueous solutions as **molarity** (the number of moles of solute per liter of solution). As a side note, we use a type of "shorthand" notation for molarity of any dissolved substance by putting the chemical formula for that substance in [square brackets]. For example, if we have 1.0 mole per liter of sodium ions, we could simply write: $[Na^+]$ = 1.0 M. (The capital letter "M" represents the unit "mol/L.")

The relationship between molarities of $[H_3O^+]$ and $[OH^-]$ that comes out from this equilibrium is shown in Eq. 9 below:

$$[H_3O^+] \times [OH^-] = 1 \times 10^{-14} \tag{9}$$

The most acidic solutions you would typically encounter have a hydronium concentration of $[H_3O^+]$ = 1.0 M, and the most basic solutions usually have a hydroxide concentration that is $[OH^-]$ = 1.0 M as well. From Eq. 9, we can draw the following for typical acids and bases. ($[H_3O^+]$ numbers are depicted at the top of the number line and $[OH^-]$ numbers are at the bottom.)

$[H_3O^+]$ →   1.0 M 1×10^{-7} M 1×10^{-14} M

1×10^{-14} M 1×10^{-7} M   1.0 M   ← $[OH^-]$

acidic solutions   *neutral*   *basic solutions*

As you can see, at one point $[H_3O^+]$ = $[OH^-]$ = 1×10^{-7} M, and the solution is **neutral**—that is, neither acidic nor basic. Also note that, as mentioned above, an increase in $[H_3O^+]$ requires a decrease in $[OH^-]$. Therefore, we don't really need to look at *both* $[H_3O^+]$ and $[OH^-]$ to determine how acidic or basic our solution is, but we can pick *just one* and use it for this purpose—usually $[H_3O^+]$.

However, the actual numbers for the molarity of H_3O^+ (or OH^-) are not "convenient" numbers to work with, but are all very small—the largest number shown is 1.0 M, and the smallest is 1×10^{-14} M (that is, 0.000 000 000 000 01 M). To make comparison of acids and bases more obvious, chemists have introduced the so-called **pH scale**. We define pH mathematically as the negative of the common logarithm of the H_3O^+ molarity (Eq. 10):

$$pH = -\log [H_3O^+] \tag{10}$$

If algebra is not your favorite subject, the sentence above and Eq. 10 may not make too much sense... ☺ While it would be nice for you to truly *understand* what common logarithms are, the bottom line is that it's a mathematical function—and a button on your scientific calculator—that allows us to transform the molarities of H_3O^+ to the following numbers (drawn below the number line):

$[H_3O^+]$ →   1.0 M 1×10^{-7} M 1×10^{-14} M

0   7   14   ← pH

Notice that the number on the pH scale is equal to the power of 10 for $[H_3O^+]$. (This is not coincidental, but is a property of the common logarithm function.) Thus, we have an acidity scale:

- Solutions with pH below 7 are acidic;
- Solutions with pH above 7 are basic;
- Solutions with a pH of exactly 7.00 are neutral!

Obviously, since this is a pH *scale*, a solution with pH = 1.0 and a solution with pH = 6.0 are both acidic, but the solution with a pH of 1.0 *is much **more acidic*** than the solution with a pH of 6.0!

In this experiment, we will measure the acidity (pH) of several solutions of acids and bases; we will also use pH measurement to help determine how much hydrochloric acid is present in a reaction mixture. In order to do this, we need to have a way of measuring pH.

pH Measurements

The simplest, quickest, and most straightforward way of measuring pH is with a **pH meter**. This is a device that uses a special electrode lowered into the solution to measure the amount of H^+ directly. The electrical signal from the electrode is fed into the "brains" of the pH meter, which then calculates the pH reading and displays it on the screen. To use the pH meter properly, you have to prepare solutions of known pH and measure them to adjust how the meter calculates pH values. A typical pH meter is shown in Figure 1.

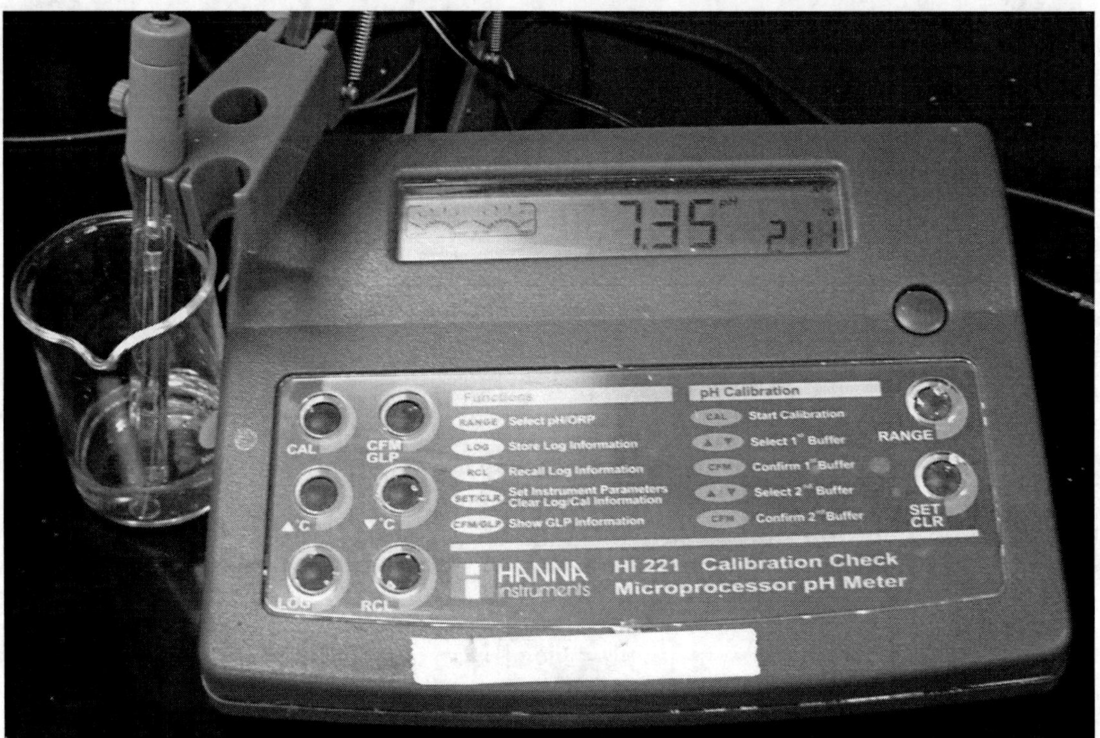

Figure 1. pH meter

Another way to measure pH, at least approximately, is by adding solutions of **acid-base indicators**. These solutions contain substances that change color depending on the acidity of their environment. Usually indicators have one color in acidic solutions and another in basic solutions, although sometimes they will have several colors in different parts of the pH range.

Some commonly used indicators are listed in Table 1, along with their colors and pH ranges. This information can be used if you have laboratory data on the color of several indicators in a particular solution to narrow down its pH range. You will do this for four solutions of acids and bases in this experiment.

<u>Table 2</u>. Commonly used acid-base indicators and their pH ranges

Indicator	"Acidic" Color	"Basic" Color	pH Range of Color Change Acid → Base
Alizarin yellow	Yellow	Red	10.1 – 11.1
Bromphenol blue	Yellow	Purple	3.0 – 4.6
Bromthymol blue	Yellow	Blue	6.0 – 7.6
Congo red	Blue	Red	3.0 – 5.0
Litmus*	Red	Blue	5.0 – 8.0
Methyl orange	Red	Yellow**	3.2 – 4.4
Phenol red	Yellow	Red	6.6 – 8.0
Phenolphthalein	Colorless	Pink (red)	8.2 – 10.0
Thymol blue	Red	Yellow	1.2 – 2.8
	Yellow	Blue	8.0 – 9.6

*Litmus usually comes as acidic "red litmus" paper, basic "blue litmus" paper, or "neutral" purple litmus paper.
**Methyl orange tends to look "quite orange" even when it "should be" yellow, if too much of it was dropped in a solution

---

EXAMPLE 1

A solution of an unknown substance gives a yellow color in phenol red, purple in bromphenol blue, and green in bromthymol blue. Adding a drop of the solution to red litmus paper and blue litmus paper does not change their color. What is the approximate pH of the solution?

Looking at Table 1 information, the following is quite obvious:

- For phenol red: yellow = a pH below 6.6
- For bromphenol blue: purple = a pH above 4.6

Thus, our pH is somewhere in this range:

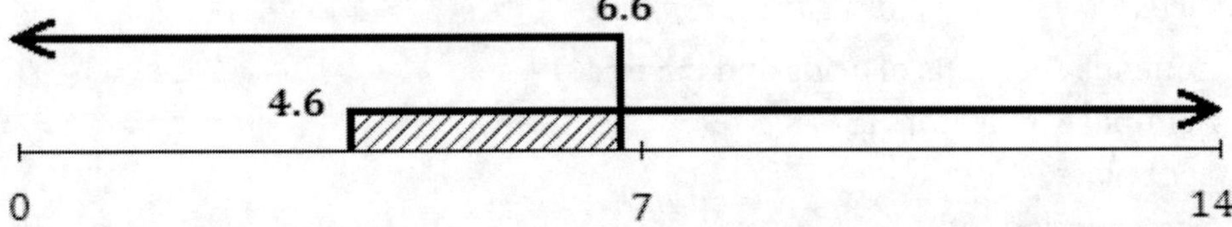

However, what does green in bromthymol blue represent? In Table 1, bromthymol blue colors are listed as yellow for pH below 6.0 and blue for pH above 7.6, and green isn't listed at all. However, **green is a mix of yellow and blue**, which means that the pH is **between 6.0 and 7.6**. This means overall, the pH of our solution should be between 6.0-6.6, as you can see from the drawing:

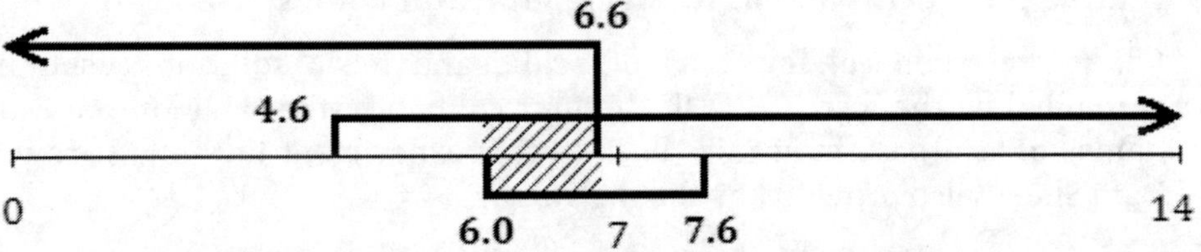

What about the red and blue litmus paper? Usually, for a solution with pH below 5.0, the red litmus will *stay* red and the blue will *change* to red. Likewise, if the solution pH is above 8.0, the red litmus will *change* to blue and the blue will *stay* blue. However, in our case, neither color changed; this is because *red litmus is already acidic by itself, and blue litmus is already basic by itself*! The drop of our solution is not acidic or basic enough to change either color; this means that our solution pH is between 5.0 and 8.0. However, this doesn't really give us any additional information, since we already knew it was between 6.0 and 6.6!

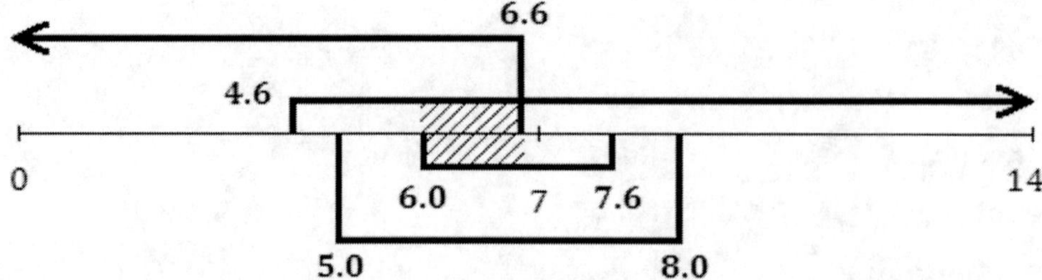

Therefore, the closest we can get to identifying the unknown solution pH is to say:

$$\text{pH} \approx 6.0 - 6.6$$

Objective

To use acid-base indicators to determine the acidity/basicity of aqueous solutions

Safety

- Wear approved **safety glasses or goggles** during the entire experiment! It is especially important when we are working with toxic, corrosive, and irritant substances.

- Aqueous solutions of acids and bases are toxic, corrosive, and can irritate or cause burns.

- Remember the location of the emergency eyewash and safety shower in case of accidents!

- Wash your hands thoroughly before leaving the laboratory.

Materials and Equipment
You will need the following for your experiment:

1. Aqueous solutions of acids and bases (4)
2. Acid-base indicators (5)
3. Spot plate
4. Glass stirring rod
5. pH meter (optional)

Procedure

I. **Estimating pH of acids and bases using indicator colors**

1. Fill in the chemical formulas of acidic and basic solutions used in this experiment (provided by the instructor) in the first column in the table on the Data Sheet. Fill in the names of acid-base indicators used in this experiment in the first row in the table on the Data Sheet (also provided by the instructor).

2. Obtain a spot plate from the instructor (shown in Fig. 2).

3. Place 1-2 drops of the first solution from the table on the Data Sheet in five of the spots on the spot plate.

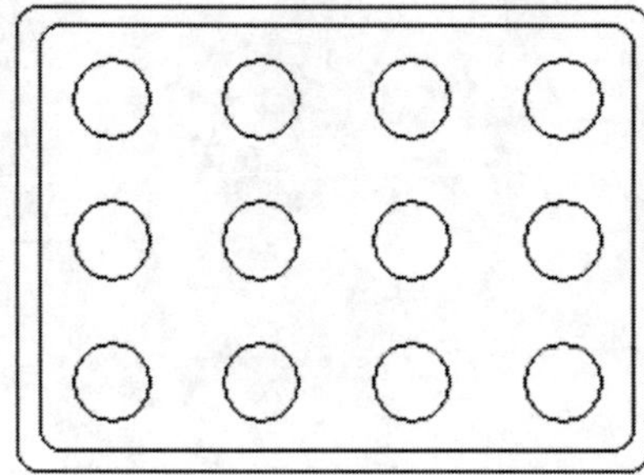

Figure 2. Spot plate

> **IMPORTANT**: *Please use the pipettes (eye droppers) provided with the solution bottles for this purpose! Do not use the same pipette for different solutions—this will lead to **contamination** and errors in your data!*

4. Add one drop of each indicator to the five samples of the solution—***one indicator per spot, NOT all five indicators in all of the spots!***

5. Write down the colors of the indicators in the samples in the table on the Data Sheet.

   (*At this point, leave the "Indicator pH" and "Measured pH" columns blank. You will fill them in later.*)

6. Use a pH meter, if provided, to record the "Measured pH" value for each solution by placing the pH meter into original bottles with solutions.

   (*The instructor may simply provide these values instead.*)

7. Discard the solutions as directed under "Clean-Up." Be careful not to get them on your hands, skin, face, or clothing!

8. Repeat Steps 2-7 for the other three solutions listed in the table on the Data Sheet.

II. Estimating pH of household solutions

9. Repeat Steps 2-7 for the other three solutions listed in the table on the Data Sheet.

Clean-Up

- Clean up any spills by using paper towels to soak in the spilled acid or base, then washing off the benchtop with plenty of water. Do this as soon as you find a spill!

- Discard the solutions from the spot plate into the collection bucket labeled "MIXED ACIDS AND BASES DISCARD," taking care not to get them on your hands. Wash off the spot plate in the sink by running water from the faucet. Dry it and continue the experiment.

- Return the clean spot plate and pH meter to the front desk.

Results

Use the data from your Data Sheet to perform estimations. Record results on the Calculations sheet.

1. Use the approach outlined in Example 1 to determine the approximate range of pH for each of the four solutions. Write down the pH range for each in the "Indicator pH" column in the table on your Data Sheet.

2. If pH meters were not used, obtain the measured pH values of the solutions from the instructor and record them in the "Measured pH" column on your Data Sheet.

<table>
<tr><td>name</td><td>section</td><td>date</td></tr>
</table>

Pre-Laboratory Assignment

1. What are specific health hazards of acids and bases? How should you protect yourself?

2. Define these terms:

   a. pH –

   b. Acid-base indicator –

   c. Molarity –

3. Using the information in Table 2, what *range* of pH values do the following solutions have?

   a. a solution of NH_3 gives a **green** color with a drop of the **thymol blue** indicator

   b. a solution of NaCl gives a **red (unchanged)** color with red litmus paper and a **blue (unchanged)** color with blue litmus paper

   c. a solution of $NaNO_2$ gives a **yellow** color with a drop of **alizarin yellow**, a **blue** color with a drop of **thymol blue**, and a change to **blue** for the **red litmus paper**

name   section   date

Data Sheet

I. Estimating pH of acids and bases:

IMPORTANT: Fill in the blank lines in the top row and leftmost column before beginning!

Indicator / Solution						Estimated pH range*	Measured pH value**

*Record your estimated range of values of pH based on the four indicators in this column.
**Report the pH meter value of pH measured in class in this column.

II. Estimating pH of household liquids:

IMPORTANT: Fill in the blank lines in the top row and leftmost column before beginning!

Indicator / Solution						Estimated pH range*	Strong/weak acid/base?**

*Record your estimated range of values of pH based on the four indicators in this column.
** Fill in this column with your determination of how strongly/weakly acidic (or strongly/weakly basic) each solution is, based on the estimated pH range

| name | section | date |

Post-Laboratory Assignment

1. You have two test tubes, each containing a colorless liquid—one of the solutions we used in lab, or possibly water! You're not sure what each one is, so you try to add a drop of phenolphthalein to each. One of them (let's call it **Solution 1**) turns hot pink, and the other one (**Solution 2**) stays colorless. Then you pour the two solutions together, and the resulting mixture turns colorless. Are Solutions 1 and 2 acidic, basic, or neutral?

   Solution 1: _______________________   Solution 2: _______________________

2. Using Eq. 10, calculate the pH of the following solutions:

   a. Milk of magnesia, $[H_3O^+] = 3.16 \times 10^{-11}$ M. *Show work.*

   b. Oven cleaner, hydrogen ion molarity 9.98×10^{-14} moles per liter. *Show work.*

   c. Tomato juice, hydronium concentration 7.94×10^{-5} M. *Show work.*

   d. Black coffee, $[H^+] = 3.16 \times 10^{-6}$ mol/L. *Show work.*

Worksheet Instructions

Use the concepts and skills you have learned in this experiment and in lecture to complete, balance, and categorize the chemical equations on the Worksheet. In each section except Part V, the first equation shows an example of how the part should be completed.

Detailed directions for each part are given below.

I. Balancing Equations and Types of Reactions

1. Balance each chemical equation (numbers 1-13) by writing coefficients on the lines provided next to chemical formulas in each one. If a coefficient is "1," leave its line blank!

2. For each equation indicate which type of reaction the equation describes (combination, decomposition, single displacement, or double displacement).

   NOTE: Some of the equations will not fit into one of these four categories! For these, the reaction type has been filled in.

II. Combination Reactions

3. Complete each reaction by writing the chemical formula of the product. Since the reactants in each reaction are a metal and a nonmetal, the product in each case will be an ionic compound. *This means your task is to come up with the **correct** chemical formula of the ionic compound resulting from the combination of the two elements! (Remember the octet rule and the "X rule" for writing formulas.)*

4. Now that you have the product formula, balance each equation.

III. Single Displacement Reactions (To be completed as instructed by your professor)

5. Complete each reaction by writing the chemical formula of the products. Here are some hints on how to do this:

   a. In each reaction, a metal will replace another metal (or replace hydrogen) from the compound. Therefore, one product will be an **element** (the metal being replaced, or H_2).

   b. The other product will be a **compound** consisting of the "new" metal as a cation, combined with the original anion from the compound reactant. Once again, *your task is to come up with the **correct** chemical formula of the ionic compound resulting from the combination of these two elements—use the "X rule" again!*

6. Now that you have the product formulas, balance each equation.

7. Rewrite the chemical equation as a complete ionic and, then, net ionic equation:

   a. Remember that only ionic compounds that are present as aqueous solutions (aq) can be written in their (dissociated/broken up) ionic form;

   b. All other components (gases, solids, liquids) will stay in their undissociated ("together") form;

   c.   Spectator ions are ones that show up *unchanged* on both sides of the equation, and they are ***dropped*** in the net ionic equation;

IV. Double Displacement Reactions (To be completed as instructed by your professor)

8. Complete each reaction by writing the chemical formula of the products. Here are some hints on how to do this:

   a.   The cations and anions "switch partners":

      i.   The cation from the first compound goes with the anion from the second
      ii.   The cation from the second compound goes with the anion from the first
      iii.   ***REMEMBER! You must still write <u>correct</u> chemical formulas of ionic products—use the "X rule!!"***

   b.   Also remember—a double displacement only occurs if one of the products is...

      o A precipitate (solid, insoluble in water)
      o A gas (seen as bubbles in solution)
      o Liquid water—$H_2O(\ell)$

   c.   If all products are aqueous solutions of ionic compounds, ***there is no reaction!*** Simply draw an "X" through the reaction arrow in the equation if that is the case!

   d.   To know whether a solid is formed, use the Solubility Chart in Table 1 above. You would check the new pairs of ions for whether they form a *soluble* compound (aq) or an *insoluble* one (s).

9. Now that you have the product formulas, balance each equation.

10. Rewrite the chemical equation as a complete and net ionic equation, as described for single displacement reactions above, in Step 7.

Worksheet

I. Balancing Equations and Types of Reactions

Balance the following chemical equations and state the reaction type:

Type

1. ____ $Pb(NO_3)_2(aq)$ + _2_ $KCl(aq) \rightarrow$ ____ $PbCl_2(s)$ + _2_ $KNO_3(aq)$ *double displ.*

$$1 - Pb - 1$$
$$2 - NO_3 - \cancel{1}\ 2$$
$$2\ \cancel{1} - K - 1$$
$$2\ \cancel{1} - Cl - 2$$

2. ____ $H_2(g)$ + ____ $O_2(g) \rightarrow$ ____ $H_2O(\ell)$ __________

3. ____ $K(s)$ + ____ $B_2O_3(s) \rightarrow$ ____ $K_2O(s)$ + ____ $B(s)$ __________

4. ____ $Na(s)$ + ____ $NaNO_3(s) \rightarrow$ ____ $Na_2O(s)$ + ____ $N_2(g)$ __________

5. ____ $C(s)$ + ____ $O_2(g) \rightarrow$ ____ $CO(g)$ __________

6. ____ $N_2(g)$ + ____ $O_2(g) \rightarrow$ ____ $N_2O_5(s)$ __________

7. _____ $C_4H_{10}(g)$ + _____ $O_2(g) \rightarrow$ ___ $CO_2(g)$ + _____ $H_2O(g)$ *combustion*

8. _____ $C_3H_8(g)$ + _____ $O_2(g) \rightarrow$ ___ $CO_2(g)$ + _____ $H_2O(g)$ *combustion*

9. _____ $NH_3(g)$ + _____ $HCl(g) \rightarrow$ _____ $NH_4Cl(s)$ __________

10. ___$Ca_3(PO_4)_2(s)$ + ___$SiO_2(s)$ + ___$C(s) \rightarrow$ ___$CaSiO_3(s)$ + ___$CO(g)$ + ___$P(s)$

          (other)

11. _____ $NH_3(g)$ + _____ $O_2(g) \rightarrow$ _____ $N_2(g)$ + _____ $H_2O(g)$ __________

12. _____ $NaOH(aq)$ + _____ $H_2SO_4(aq) \rightarrow$ _____ $Na_2SO_4(aq)$ + _____ $HOH(\ell)$ __________

13. _____ $C(s)$ + _____ $SO_2(g) \rightarrow$ _____ $CS_2(\ell)$ + _____ $CO(g)$ *(other)*

          ____

II. Combination Reactions

Complete and balance the chemical equations for the following combination reactions.

14. __*4*__ Na(s) + ____ O$_2$(g) → __*2*__ ___*Na$_2$O*___(s)

$$4\ \cancel{1} - Na - \cancel{2}\ 4$$
$$2 - O - \cancel{1}\ 2$$

15. ____ Al(s) + ____ S(s) → ____ _______________(s)

16. ____ Cs(s) + ____ N$_2$(g) → ____ _______________(s)

17. ____ Mg(s) + ____ Cl$_2$(g) → ____ _______________(s)

18. ____ Rb(s) + ____ P(s) → ____ _______________(s)

Complete and balance the chemical equations for the following single displacement reactions. Write complete and net ionic equations for each.

19. ____ Zn(s) + ____ CuCl₂(aq) → ____ ____**Cu**____(s) + ____ ____**ZnCl₂**____(aq)

$$1 - Zn - 1$$
$$1 - Cu - 1$$
$$2 - Cl - 2$$

complete ionic: **Zn(s) + Cu²⁺(aq) + 2 Cl⁻(aq) → Cu(s) + Zn²⁺(aq) + 2 Cl⁻(aq)**

net ionic: **Zn(s) + Cu²⁺(aq) → Cu(s) + Zn²⁺(aq)**

20. ____ Ca(s) + ____ HCl(aq) → ____ H₂(g) + ____ ______________(aq)

complete ionic:

net ionic:

21. ____ Mg(s) + ____ AgNO₃(aq) → ____ __________(s) + ____ ______________(aq)

complete ionic:

net ionic:

III. Single Replacement and Double Displacement Reactions

Complete and balance the chemical equations for the following double displacement reactions. Rewrite the equation as a complete ionic and net ionic equation.

22. __*3*__ KOH(aq) + ____ H_3PO_4(aq) $\rightarrow$ ____ __*K_3PO_4*__(aq)+ __*3*__ ____*HOH*____(ℓ)

$$3 \; \cancel{1} - K - 3$$
$$3 \; \cancel{1} - OH - \cancel{1} \; 3$$
$$3 - H - \cancel{1} \; 3$$
$$1 - PO_4 - 1$$

complete ionic:
3 K$^+$(aq) + 3 OH$^-$(aq) + 3 H$^+$(aq) + PO$_4$$^{3-}$(aq) $\rightarrow$

3 K$^+$(aq) + PO$_4$$^{3-}$(aq) + 3 HOH($\ell$)

net ionic: **3 OH$^-$(aq) + 3 H$^+$(aq) $\rightarrow$ 3 HOH(ℓ)**

23. ____ NaCl(aq) + ____ $AgNO_3$(aq) $\rightarrow$ ____ ____________(s)+ ____ ____________(aq)

complete ionic:

net ionic:

24. ____ $FeCl_3$(aq) + ____ NaOH(aq) $\rightarrow$ ____ ____________(s)+ ____ ____________(aq)

complete ionic:

net ionic: